Lecture Notes on Vertebrate Zoology

Lecture Notes on Vertebrate Zoology

Ronald Pearson and John N. Ball

MA, PhD
Reader in Zoology
Liverpool University

BSc, PhD
Professor of Zoology
Sheffield University

A Halsted Press Book

John Wiley & Sons

NEW YORK — TORONTO

First published 1981

Published in the U.S.A. and Canada
by Halsted Press, a Division of
John Wiley & Sons, Inc, New York

Library of Congress
Cataloging in Publication Data

Pearson, Ronald George.
    Lecture notes on vertebrate zoology.

    "A Halsted Press book."
    Includes index.
    1. Vertebrates. I. Ball, John N., joint author.
II. Title.
QL605.P34   596   80–29672

ISBN 0–470–27143–4

Printed in Great Britain

Contents

Preface

This book is intended to serve a similar purpose to *Lecture Notes on Invertebrate Zoology* by M.S. Laverack and J. Dando. However, the vertebrates, as members of a single phylum, clearly form a more cohesive unit than do the thirty or so invertebrate groups. It is partly for this reason that we have not considered it appropriate to use the fully 'note' format employed by Laverack and Dando in their treatment of the extreme diversity found amongst invertebrates. Furthermore, the teaching of vertebrate biology varies greatly between institutions and we accept that no one textbook will satisfy all the requirements of all undergraduate students. We have, therefore, attempted to provide in this book a balanced account of vertebrate biology and evolution which will form a background for various kinds of elementary course and thereby provide a sound basis for later specialization in various directions.

We have deliberately placed greater emphasis on anatomy than has become usual in recent years, believing that it is desirable to correct what could become a swing away from the thorough grasp of structure that is surely an essential prerequisite for understanding physiology, evolution, behaviour, and all those other aspects of biology for which morphological and anatomical information can so usefully act as a conceptual framework. Another feature is the amount of attention given to the nervous and muscular systems, too often skimped in undergraduate teaching. In these trends we have followed Laverack and Dando in aiming, as they put it, 'to release the teacher from the drudgery of simply reproducing anatomy' and to create 'a platform for profitable expansion on innumerable other topics'.

We hope that our platform is soundly enough constructed to be useful to both teacher and student, and to permit expansion in a variety of ways. If the book encourages students to appreciate the full panorama of vertebrate biology, we will feel that our intentions have been achieved.

R.P., J.N.B.

Acknowledgments

We wish to express our appreciation to our colleagues in the Zoology Departments of Liverpool and Sheffield Universities for countless stimulating discussions over many years. We would also like to thank Miss Anita Callaghan and Mr Bryan V. Lewis of the Department of Zoology, Liverpool, for, respectively, typing the manuscript and assisting with the illustrations. Needless to say they are in no way responsible for any inadequacies which the book may have.

We are very grateful to the following authors and publishers for generously granting us permission to use the illustrations cited: Professor L.B. Halstead for Fig. 3.6 from *The Pattern of Vertebrate Evolution*, published by Oliver and Boyd; Professor A.I. Dagg for Fig. 6.15 from *Running, Walking and Jumping*, Wykeham Publications; Professor A.S. King and Dr. J. McLelland for Fig. 8.1 from *Outlines of Avian Anatomy*, Baillière and Tindall; Dr C.H. Tyndale-Biscoe for Fig. 9.18 from *Life of Marsupials*, Edward Arnold; Dr. C.J. Pennycuick for Fig. 8.4 from *Animal Flight*, Edward Arnold; Masson, Editeur, s.a., Paris, for Figs 2.7, 2.18, 8.28, 10.6–10.10 and 10.20 from the *Traité de Zoologie*; Cambridge University Press for Figs 3.12, 4.11 and 4.12 from Alexander R. M^c, *The Chordates*, and Fig. 3.22 from Fridberg G. & Bern H.A. (1968). The urophysis, *Biol. Rev,* **43**; The University Tutorial Press for Figs 3.17 to 3.19 and 3.24 from *Animal Biology* by Grove A.J. & Newell G.E; Academic Press Inc for Fig. 3.20 from Jasinski A., *Gen. Comp. Endocrinol.* suppl. 2, pp. 510–521, for Fig. 7.8 from Kochva E., *Biology of the Reptilia*, C. Gans (ed.) Vol. 8, for Fig. 3.21 from Copp D.H., in *Fish Physiology*, Hoar W.S. & Randall D.J. (eds) Vol. 2., and together with the *Journal of Comparative Neurology*, for Figs 4.18 and 7.14 which are modified versions of Figures in Heric T.M. & Kruger L. (1965) *J. Comp. Neurol.* **124** and Schwassman H.O. & Kruger L. (1965) *J. Comp. Neurol.* **124** as used by Pearson R., *The Vertebrate Brain*, Academic Press Inc; W.H. Freeman and Co., for Fig 6.1 from Goin C.J., Goin O.B. & Zug G.R. (1978) *Introduction to Herpetology*, and Fig 9.19 from Eckert R. & Randall D. (1978) *Animal Physiology*; Sidgwick and Jackson for Fig 7.11 from Carter G.S., *Structure and habit in vertebrate evolution*; both Macmillan Publishing Co., Inc., and Kendall/Hunt Publishing Co., for Figs 3.15, 6.13, 9.4, 9.5, 9.14 from Holmes, *Manual of comparative anatomy*; Alan R. Liss Inc., for Fig. 8.7 from *J. Comp. Morphol.* **142**; Queensland University Press for Fig. 9.10 from May A.D.S., *Anatomy of the sheep*; Holt, Rhinehart and Winston for Fig. 9.11 from Villee *et al.*, *General Zoology*; W.B. Saunders and Co., for Fig. 9.12 from Miller H.E., Christensen G.C. & Evans H.E., *Anatomy of the dog*; McGraw Hill Book Co., for Figs 6.19 and 6.20 from Weichert C.K., *Representative Chordates*; Kindler Verlag Gmbh, for Fig. 8.27 from Grzimek's *Animal Life Encyclopaedia*.

1 · Introduction

GENERALITIES

Vertebrates comprise a series of adaptive radiations from ancestral forms whose fossil remains first appear in rocks dating from some 450 million years ago. The precise invertebrate origins of the vertebrates remain unknown but they clearly shared a distant common ancestor with the echinoderms, a more recent one with the other non-vertebrate chordates, and themselves exhibit variations on a common plan. It is a study of these variations which forms the basis for courses in vertebrate zoology. This book considers each class in turn. Although a relatively uniform approach is used in each case some topics only relate to certain classes, and others, which relate to more than one class, are dealt with in detail on the first occasion. For information on the related phyla of invertebrates the reader is referred to Laverack & Dando (1979).

GENERALIZED STRUCTURE

i Fig. 1.1 is an exploded stereogram of a generalized vertebrate body. In general terms the body is supported throughout by an all-pervading framework of connective tissue which binds the organ systems together. There is always, at least during embryonic development, a notochordal rod which underlies the tubular, dorsal central nervous system. The central and peripheral nervous systems which are of ectodermal origin, together with the dorsal musculature and kidney ducts, which are of mesodermal origin, are all primitively arranged in a metamerically segmented pattern. This segmentation may be obscured in the adult but it is always present in the embryo. Broadly speaking, all the mesodermal structures arise during ontogeny as paired outgrowths from the embryonic gut, or archenteron,

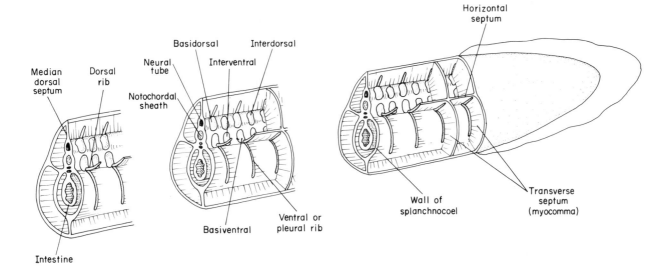

Fig. 1.1 An exploded stereogram showing the connective-tissue systems in an idealized craniate and the disposition of notochord, nervecord, etc. (After Goodrich.)

and the archenteric walls then rejoin beneath them and complete the lining of the gut.

ii Each coelomic outgrowth derived in this way, whether hollow or solid, becomes divided into dorsal and ventral moieties. The dorsal moiety subsequently gives rise to the segmentally arranged muscles, myotomes, and to the sclerotomes that give the bony or cartilaginous axial skeleton. The ventral moieties are never truly segmental and soon join to form a continuous longitudinal coelomic chamber, the splanchnic coelom, which surrounds the viscera. The intermediate mesoderm in the region linking these dorsal and ventral

components is the source of the primitive kidney structures.

iii The paired origin of the coelomic derivatives leaves the body divided into left and right halves by a median longitudinal, vertical septum. This septum persists as a suspension for the gut.

iv The sclerotome gives rise to vertebral components, around or near the notochord, and to ribs. These last extend out from the sides of the vertebrae and usually lie between the dorsal and ventral muscles but fish possess two sets of ribs.

Table 1.1 The vertebrate succession through geological time together with the predominant known vegetation and the ages as determined by radiometric methods.

Era	Period	Age from beginning (million years B.P.)	Type of vegetation	Vertebrate succession
CENOZOIC	Quaternary	2.5	Modern	Mammals
	Tertiary	65	Modern	
MESOZOIC	Cretaceous	136	Gymnosperms dominant in the Lower Cretaceous	Diapsid reptiles, plagiaulacid and ptilodontid multituberculates, marsupials, and placentals
	Jurassic	195	Luxuriant forests of gymnosperms and ferns	Diapsid reptiles, triconodonts, symmetrodonts and pantotheres
	Triassic	225	Sparse desert flora giving way to luxuriant forests of gymnosperms and ferns.	Diapsid reptiles, tritylodonts and ictidosaurs
PALEOZOIC	Permian	280	Tall swamp forests with early gymnosperms, *Calamites* and ferns giving way to desert flora of conifers and Bennettitales	Cotylosaurs, pelycosaurs and therapsidans
	Carboniferous	345	Early gymnosperms, tree lycopods and ferns	Cotylosaurs and labyrinthodonts
	Devonian	395	Herbaceous marsh plants, e.g. *Psilophyta* and *Zosterophyllum*. *Rhynia* vegetation in Middle Devonian	Acanthodians, antiarchs and ichthyostegalians
	Silurian	440	Marine algae	Agnathans
	Ordovician	500	Marine algae	Agnathans
	Cambrian	570	Marine algae with some evidence of land plants	

The environment and time-scale

The time-scale over which vertebrate evolution has occurred is summarized in Table 1.1. It is, however important to emphasize the great changes which have occurred in the environment during this period of some 450 million years. The continents have split, separated and reassembled. Climatic deteriorations and ameliorations have come and gone. In particular we have extensive evidence that glaciation occurred around the end of the Ordovician, during the Carboniferous and Permian, and during the last two million years. Complementarily we have the immense coal measures that record swamp conditions in the Carboniferous. These are not isolated events but reflect the persistent environmental variations to which vertebrates, and all other groups, have been exposed. Such changes undoubtedly greatly influenced the tempo of evolution by varying the precise nature and intensity of the selective factors acting at given moments of world history.

FURTHER READING

Suggestions for further reading follow many of the subsequent chapters. We provide here a list of textbooks that can be consulted for the purposes of essay writing, etc.

ALEXANDER R.McN. (1975) *The chordates.* Cambridge University Press, Cambridge.

CARTER G.S. (1967) *Structure and habit in vertebrate evolution.* Sidgwick and Jackson, London.

GRASSE P.P. (ed.) *Traité de Zoologie.* Masson, Paris. In progress.

GRZIMEK B. (ed.) (1974) *Grzimek's animal life encyclopaedia.* Van Nostrand, Berkshire.

HILDEBRAND M. (1974) *Analysis of vertebrate structure.* John Wiley and Sons, New York.

HOLMES E.B. (1975) *Manual of comparative anatomy.* MacMillan, London.

McFARLAND W.N., POUGH F.H., CADE T.J. & HEISER J.B. (1979) *Vertebrate life.* Collier MacMillan, New York.

OLSON E.V. (1971) *Vertebrate paleozoology.* Wiley-Interscience, New York.

ROMER A.S. (1967) Major Steps in Vertebrate Evolution. *Science,* **158**, 1629–37.

ROMER A.S. (1971) *Vertebrate paleontology,* (3rd ed.). University of Chicago Press, Chicago.

ROMER A.S. & PARSONS T.S. (1977) *The vertebrate body.* W. B. Saunders, New York.

SCIENTIFIC AMERICAN READINGS: (1974) *Vertebrate structure and function.* W. H. Freeman & Co, San Francisco.

STAHL B.J. (1974) *Vertebrate history: Problems in evolution.* McGraw Hill, New York.

WEBSTER D. & WEBSTER M. (1974) *Comparative vertebrate morphology.* Academic Press, London and New York.

WILLIAMS W.D. (1972) Parker and Haswell's *Textbook of zoology.* (8th ed.) 2 vols. Macmillan, London.

YOUNG J.Z. (1980) *The life of the vertebrates,* (3rd ed.). Oxford University Press, Oxford.

2 · The Superclass Agnatha

N.B. The relationships of the hagfish are controversial. Some authors affiliate it to the pteraspidomorphs, whilst others deny its relationships to known fossil forms.

INTRODUCTION

The cyclostomes are the most primitive living vertebrates and their fossil relatives, the ostracoderms, were the first animals with backbones to appear in the fossil record. The absence of jaws is itself a distinctive characteristic serving to distinguish the agnathans from all the jawed vertebrates or gnathostomes. However, they possess a number of other unique features:
  their sucking or rasping mouths;
  the unpaired olfactory organ of cephalaspidomorphs;
  the absence of paired fins;
  the absence of both pectoral and pelvic girdles;
  the presence of only one or two semicircular canals in the inner ear.
They have well-developed anal fins but bending the body by serial contraction of the myotomes provides the principal motive force. The presence of paired fin-like structures in some fossil forms, and the presence of paired olfactory capsules in pteraspidomorphs, coupled with the fact that the unpaired olfactory organ of cyclostomes is associated with two olfactory nerves, implying a dual origin during phylogeny, demonstrates that the modern genera combine primitive and specialized characteristics.

GENERAL FEATURES AND THE LIFE CYCLE OF MODERN FORMS

i At least two distinct lineages are clearly represented by the fossil forms that are known collectively as ostracoderms. The Pteraspidomorpha, Diplorhina or Heterostraci, possessed two nostrils and are therefore envisaged as approximating most closely to the ancestral stock. The Cephalaspidomorpha or Monorhina, including both the Osteostraci and Anaspida, have a single nostril (Fig. 2.1 and 2.2). The anaspids, in particular, have been viewed as representing a form which is closely similar to that of the putative ancestors of modern genera.

ii Known from the Ordovician, Silurian and Devonian periods the three groups underwent extensive adaptive radiations during the Upper Silurian and Lower Devonian only to decline rapidly thereafter. Apart from lineages which may be ancestral to the cyclostomes, the majority became extinct by the early Carboniferous. They exhibit a variety of buccal adaptations. These suggest that their methods of feeding ranged from microphagous ciliary feeding, comparable with that in Amphioxus or the ammocoete larva of lampreys, to swallowing larger food masses, as do the majority of vertebrates. Spoonlike adaptations were presumably for shovelling up bottom detritus which was then exposed to pharyngeal filtration mechanisms.

iii The two groups of living cyclostomes are variously ascribed to distinct orders or superorders by different workers. In broad terms they are the Petromyzontoidea, or lampreys, and the Myxinoidea, or hagfishes (Fig. 2.3). The latter comprises three genera—*Myxine*, *Paramyxine* and *Eptatretus*. They are marine forms that either burrow into dead or dying fishes, or eat marine invertebrates. Their lateral eyes are functionless, rudimentary organs, and they lack a pineal eye, but horny teeth are well developed.

Fig. 2.1 Lateral view of a pteraspidomorph.

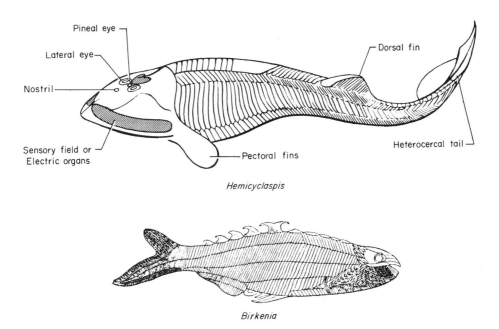

Fig. 2.2 Lateral views of the osteostracan *Hemicyclaspis* and the anaspid *Birkenia*.

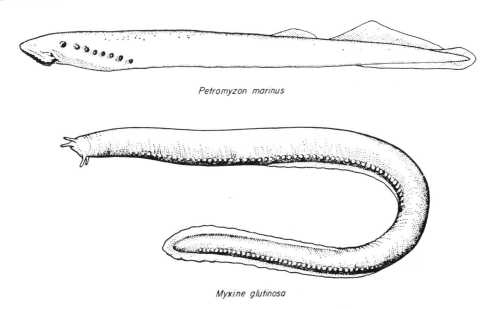

Fig. 2.3 *Petromyzon marinus*, the marine lamprey, and *Myxine glutinosa*, a hagfish.

iv The Petromyzontoidea includes eight genera. The life cycle consists of a filter-feeding larval phase, the ammocoete larva, possessing many anatomical features reminiscent of Amphioxus, and a fish-like adult which is parasitic on fishes. This adult phase is achieved after a larval period of some three to seven years spent in the mud of freshwater lakes and streams. Some species then migrate to the sea whereas others remain in freshwater environments. In either case, fully sexually mature adults migrate to freshwater streams for spawning. They have an anterior sucker bounded by a series of lips with which they attach themselves to the host fish. They then rasp away at the body surface of the fish by means of their tongues. Apparently they live primarily upon the host's blood and, having fed, detach themselves and swim away.

THE AMMOCOETE LARVA

i The body of the ammocoete is metamerically segmented. During development the coelom and mesoderm originate from dorsolateral pouches which become nipped off from the archenteron, or embryonic gut, during the gastrula stage. The notochord arises as a third outgrowth in the mid-dorsal line at a slightly later stage. This method of forming the coelom and mesoderm is referred to as enterocoelic. Each somite then becomes differentiated into a dorsal myocoel and a more ventral splanchnocoel. The former is segmentally organized but the latter is not, and instead forms the continuous cavity surrounding the gut. On each side an intermediate component of the coelom, the nephrocoel, retains its opening into the splanchnocoel, the coelomostome, and gives rise to the kidney rudiments.

ii The median wall of the original myocoel differentiates in three ways. The majority of it forms the segmental muscle blocks, myotomes, which grow out into the original cavity and more or less occlude it. Growing ventrally they then attain a position in the body wall outside the splanchocoel. A small ventral component of the original wall grows between the myocoel and the notochord, is known as the sclerotome, and subsequently develops into skeletal tissue. A further dorsal part is associated with skeletal tissue in the fin.

iii In the cranial region the early segmentation is obscured later by the development of the organs in the head. In the ammocoete, as in the embryos of all other vertebrates, the following cranial segments can be detected. The first segment lies in front of the mouth, is 'premandibular', and its myotomal derivatives form those extrinsic eye muscles that are innervated by the third, or oculomotor, nerve. These are the inferior oblique, the superior, inferior and internal rectus muscles. Behind this segment, but in front of the ear, there are two further segments (2 and 3) whose

myotomes also contribute to the eye musculature. As the eyes of ammocoetes are small, lying below the skin, these myotomes are also small. Segments 4 and 5 are in the ear region and their myotomes are displaced backwards but behind the otic capsule the sixth and subsequent segments form a regular sequence.

Gill clefts form on each side in the visceral plate region of segments 3–10, but those in segment 3 soon close. There are, therefore, some seven functional gill clefts which grow backwards behind their segments of origin during development. The ventral parts of their homologous myotomes grow back with them and displace the ventral parts of the myotomes in the segments behind. These ventral myotomal components then approach the mid-ventral line and form the hypobranchial musculature.

iv The cranial nerves are summarized in Table 2.1. The ventral nerve roots of segments 1–3, the oculomotor (III), trochlear (IV) and abducens (VI) nerves, supply the extrinsic eye muscles formed from the myotomes of these pre-otic segments. The ventral roots of segments 4 and 5 do not develop, although the myotomes are present and are innervated by more posterior roots. The ventral roots of segments 6–10 are typical nerves to the myotomes and those of 11–18 supply the hypobranchial musculature.

Table 2.1 The segmental organization of the cranial region in the ammocoete larva.

Segment	Dorsal root	Ventral root	Muscles served
1	Profundus (usually referred to as V_1)	Oculomotor III	Inferior oblique, superior, inferior and internal recti
2	Trigeminal (V_2 and V_3)	Trochlear or pathetic IV	Superior oblique
3	Facial VII	Abducens VI	External rectus
4	Glossopharyngeal IX	Absent	Displaced backwards behind the otic region
5	Vagus	Absent	
6–10	Vagus	Normal roots supplying myotomal derivatives of segments 4–10	Myotomes

The profundus nerve is apparently a branch of nerve V in most vertebrates. In the ammocoete it is free. It is the dorsal root of the first segment and supplies sense organs at the front of the head. The dorsal root of the second segment is the trigeminal nerve (V), that of the third segment is the facial (VII) which innervates cranial lateral-line sense organs. The glossopharyngeal (IX) is the dorsal root of the fourth segment and innervates the first functional gill slit. The dorsal roots of segments 5–10 combine to form the vagus nerve (X) which not only provides the innervation for the remainder of the gill clefts but also has a branch which innervates the lateral line sense organs on the side of the body. It can be seen that motor neurons travel in dorsal roots in the cranial region which therefore retains an ancient organization comparable with the trunk region in Amphioxus. The basic structure involves modification of an underlying, repetitive, metameric segmentation (see later Table 3.1).

v The ammocoete larva is a ciliary or microphagous feeder. Two ectodermal flaps on the sides of the head grow ventrally and form an oral hood. Around the mouth there are some papillae which both prevent large particles from entering the pharynx and bear sense organs that test the inhalant water. The internal oral cavity is a stomodaeum lined by ectoderm and two muscular flaps—the velum—project backwards at the connection between the oral cavity and the pharynx. Rhythmical contractions of both these flaps and the pharyngeal wall result in a current of water passing into the pharynx and then out to the exterior via the gill clefts. In the mid-ventral line of the pharynx there is a ciliated groove which divides into two peripharyngeal bands that pass dorsally. Within this ciliated groove, and at the level of the second functioning gill slit, there is the opening of the endostyle, an organ which secretes mucus by which food particles are trapped and passed back into the oesophagus.

Between the pharyngeal wall and the external openings of the gill clefts lie the gill chambers (see Fig. 2.4) whose organization is more similar to that of fishes than are those of adult lampreys. Gill filaments develop on both the anterior and posterior walls as a series of horizontal folds. Above and below each filament a series of secondary folds is organized transversely to it and it is here that gaseous exchange occurs.

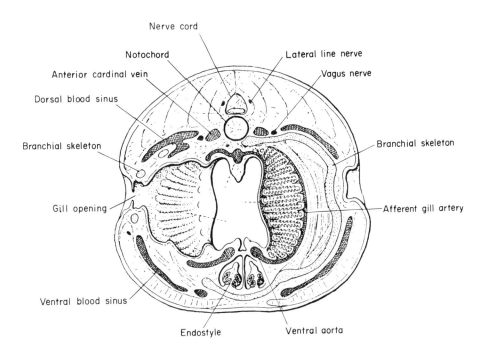

Fig. 2.4 Transverse section of the branchial region of an ammocoete larva passing through a gill bar on the right and a gill opening on the left.

THE INTEGUMENT

i Sections cut perpendicular to the surface reveal the structure of the body wall. In the pteraspidomorphs there was a superficial layer of dentine which tended towards enamel on the surface. Below this there was a reticular layer of bone. In the osteostracans an analogous dentine layer was penetrated by very fine canals, similar to those in the dentine of modern vertebrates and opening onto the surface by way of minute pores (Fig. 2.5). The superficial layer was also penetrated by much larger pores. It has been suggested that mucus was secreted through these in view of the great importance of mucus as both an osmotic control and a general protection in modern hagfishes. The layer was also very varied in its development. In some genera it is a prominent feature whilst in others it can either be restricted to the surface of tubercles or be completely absent. It was underlain by a layer containing large diameter anastomosing canals which contained a network of blood vessels, and, below this again, was a basal layer of laminated bone containing many spaces for cells and some large canals through which subcutaneous blood vessels coursed.

ii The skin of *Myxine* consists of a hypodermis rich in fat cells, a dermis with bundles of collagenous fibres, and an epidermis. These are, respectively, 240 μm, 90μm and 90 μm thick. The epidermis is unique. There are five types of cells—undifferentiated basal cells, small mucous cells, large mucous cells, thread cells and sensory cells (Fig. 2.6). Undifferentiated basal cells and small mucous cells comprise the bulk of the inner and outer halves of the epidermis respectively, and their numbers greatly exceed those of other types. Secretory canaliculi within the apical regions of the mucous cells

give a striated appearance but it is the thread cells that are the most conspicuous individual units. They have a large central mass of dense granules and peripheral spiral threads which lie within an aqueous matrix. When discharged these form part of the slime.

iii In the midline the skin is produced into the fins—two on the back and one round the tail in lampreys. These are supported by fin rays. In the cephalaspidomorphs the one or two fins were a continuation of a dorsal crest and this was also carried over the tail to form a caudal fin.

SKELETAL SYSTEM

(a) Cephalic region

i In the pteraspidomorphs and osteostracans the anterior region of the body was encased within the heavily ossified carapace. In pteraspidomorphs this formed a

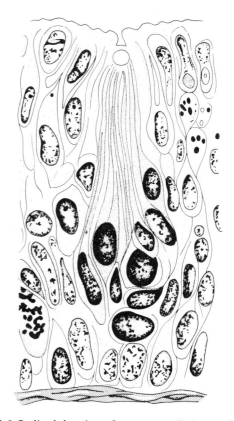

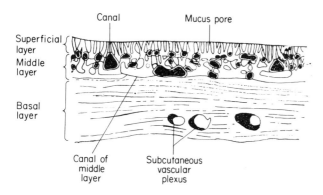

Fig. 2.5 Section through the carapace of an osteostracan.

Fig. 2.6 Stylized drawing of secretory cells in the skin of *Myxine*.

relatively immobile unit of variable appearance. It was composed of large plates, which may have originated by the fusion of scales, and could be either rounded or pointed. Such distinct plates were lacking in osteostracans. However, in both cases the carapace enclosed both the cranial and branchial regions and, in the osteostracans, where it was more strongly flattened on the ventral side, it also encased most of the abdominal region. In contrast, the anterior of anaspids lacked a coherent carapace and was covered by small plates.

ii The internal skeleton of pteraspidomorphs was not ossified and is poorly known. The osteostracans had an ossified cranial skeleton whose composition contrasts with that of most vertebrates. Instead of a separate braincase and jointed branchial arches the dermal shield was underlain by a single unified structure. It has been suggested that this represents a primitive undifferentiated condition and that the establishment of separate

units followed later, although it is probable that an early metameric organization occurred during development. The internal skeleton of anaspids is presumed to have been cartilaginous.

iii Modern cyclostomes have a skeleton of uncalcified cartilage which differs from that in other vertebrates in having relatively little matrix. The incomplete neurocranium which protects the sense organs and brain is similar in both the ammocoete larva and the adult lamprey—although more complete in the latter—but the splanchnocranium changes in association with the transition from larval microphagy to the adult parasitic way of life. In broad terms the neurocranium develops from paired rudiments (Fig. 2.7). These comprise paired trabeculae and olfactory capsules in front; parachordals on either side of the notochord; lateral basitrabecular apophyses and otic capsules. The trabeculae of gnathostomes have long been considered as representing the remnants of a gill arch in segment one. In the adult cyclostome they are overlain by the lateral cranial wall and a cranial roof forms between the otic capsules, but the fact that both the glossopharyngeal (IX) and vagus (X) nerves leave the brain behind the caudal limits of the skull shows that the occipital region is essentially incomplete by comparison with that in gnathostomes.

iv The splanchnocranium is a continuous cartilaginous structure known as the branchial basket (Fig. 2.8). Slender, irregular, cartilaginous bars lie external to the gill pouches and support the gill region but there are several reasons for not homologizing them with the typical visceral arches of gnathostomes, and such homologies remain controversial. Additional, anterior, dorsal and ventral cartilages provide support for the rasping tongue and sucker.

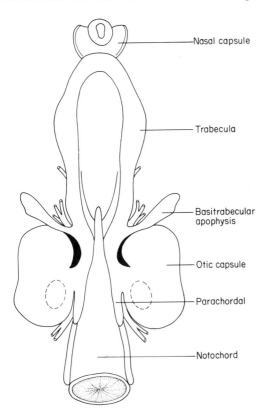

Fig. 2.7 Ventral view of cranial skeletal rudiments in *Lampetra*. (After Parker.)

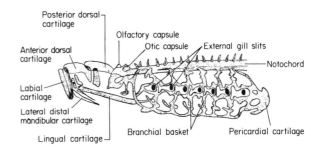

Fig. 2.8 Lateral view of chondrocranium and branchial basket of *Petromyzon*. (After Parker.)

(b) Axial skeleton

i As noted above, the internal skeleton of pteraspidomorphs was not ossified and is poorly known. In osteostracans ossification took place in association with pre-existing cartilage. Perichondral ossification left the cartilage persisting within a bony sheath, whilst endochondral ossification transformed it into bone. The majority of genera had a pair of pectoral fins that were probably supported by fin rays and the tail was heterocercal with the vertebral elements in the dorsal lobe. In contrast the pteraspidomorphs had hypocercal tails and, during swimming, the upward movement, resulting from the more flexible ventral lobe, depressed the front end of the body (see page 32). The tail of anaspids such as *Birkenia* tilted downward to give a reversed heterocercal tail. Amongst living forms this is only known in the ammocoete larva and not in any other vertebrate.

ii The axial skeleton of modern cyclostomes largely comprises the persistent notochord with its vacuolated cells. Its fibrous sheath is continuous with the connective tissue septa which separate the myotomes and it functions as an elastic rod flexed laterally by myotomal contractions. Various associated perichordal structures can be cartilaginous in adult petromyzontids but are fibrous or membranous in myxinoids. In each segment two pairs of peglike structures lie on each side of the notochord. The anterior pair is in front of the ventral nerve roots, the posterior in front of the dorsal root. They are analogous, if not fully homologous, with the interdorsal and basidorsal elements of other classes, extend dorsally, and, like the neural arches of gnathostomes, partially surround the spinal cord. Similar cartilages in the tail region extend ventrally to form haemal arches. Very slender fin rays support the median fins.

MUSCULAR SYSTEM

i As cyclostomes lack paired fins the musculature of the trunk and tail has a very straightforward plan. The muscle blocks of lampreys are metamerically arranged myomeres or myotomes that are disposed almost vertically but curve forward slightly in both the dorsal and ventral regions giving a somewhat W-shaped structure. The component muscle fibres are not inserted onto the axial skeleton but onto the tough, connective tissue partitions called myosepta or myocommata which separate adjacent myotomes. The scale rows of osteostracans and anaspids are thought to reflect a com-

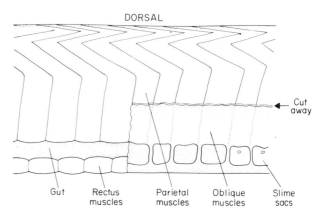

Fig. 2.9 Lateral view of body musculature in *Myxine*. The oblique muscles underlying the parietal ones are shown on the right. (After Cole.)

parable myotomal plan. In adult hagfishes there is rather more differentiation. On each side there is a metamerically organized dorsolateral parietal muscle and a continuous ventrolateral sheet forming the oblique muscle. A mid-ventral rectus muscle completes the picture (Fig. 2.9).

ii The musculature of the cranial region is, of course, modified in association with feeding and sensory functions. In lampreys the somitic mesoderm gives rise to the extrinsic eye muscles, and eight other muscles, associated with the anterior cartilages, are involved in attachment to the prey. In hagfishes the eyes are rudimentary, eye muscles are lacking, and the greater part of the cranial musculature is involved with the dental plate but a number of smaller muscles move the tentacles, lips and velum.

THE DIGESTIVE SYSTEM

(a) The buccal region

The mouth of ostracoderms was situated at the anterior end of an orobranchial chamber and varied in shape from a circular opening to a transverse or longitudinal slit. The mouth of the microphagous ammocoete larva has already been described (page 7). When attached to its prey, the mouth of the adult lamprey is round, and numerous deciduous, horny teeth are distributed on the lips and rasping tongue. That of *Myxine* is horseshoe shaped, bordered by two pairs of tentacles, and bears some 33 permanent, horny teeth. A single, median, palatal tooth is situated dorsally, points backwards, and

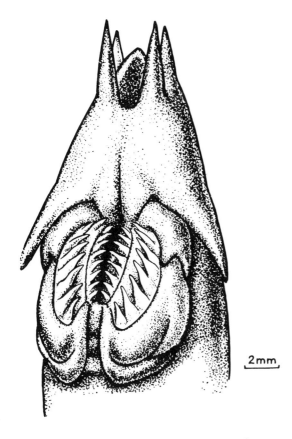

Fig. 2.10 Head of *Myxine* with dental plate bearing lingual teeth extruded.

is attached to the overlying palatal cartilage. The lingual teeth (Fig. 2.10) are associated with the dental plate cartilage which is moved by protractor and retractor muscles. Two salivary glands occur in adult lampreys but are absent from both the ammocoete larva and myxinoids.

(b) The alimentary canal and feeding

i The pharynx of *Petromyzon* is a blind pouch from which the seven internal gill slits open on each side. The blood, mucus and tissue fragments upon which the animal feeds pass out of the buccal cavity and go down the oesophagus. A valve-like structure, the velum, at the anterior end of the pharynx, prevents them entering the respiratory system and also prevents any return of water from the respiratory system which would dilute the blood. As the lamprey prepares to feed, the buccal funnel is collapsed and any water already in it is forced

over the tongue and into the mouth cavity or its overlying extension, the hydrosinus. The tongue is then used to block the passage between the buccal funnel and mouth cavity, whilst the suctorial disc is arched, thereby increasing the volume and degree of suction of the buccal funnel. Water within the hydrosinus and oral cavity is forced past the velum and into the pharynx. The tongue is now free to move back and forth. Blood flowing from the wound produced by this rasping movement passes into the buccal cavity and thence to the oesophagus.

ii The oesophagus continues posteriorly as a straight tube and joins the intestine which extends directly to the anus, although slight enlargements at either end may be considered as a stomach and rectum respectively. A conspicuous fold in the intestinal wall, the typhlosole or spiral valve, projects into the intestinal lumen and increases the surface area available for absorption. Hagfishes also lack a true stomach but the transitional region between the fore- and hindgut is demarcated by muscle fibres. In all cases the rectal region, at least, is suspended by a dorsal mesentery that results from the apposition of the coelomic components of each side.

(c) The liver and exocrine pancreas

In *Petromyzon* the single-lobed liver surrounds the 'stomach' region. Its anterior end is concave and partially surrounds the pericardium. In *Myxine* the liver has two lobes with the posterior being the larger—that of ammocoetes is similar. The bile duct, present in the latter, atrophies at metamorphosis and the adult lamprey has neither a bile duct nor a gall bladder.

No organ comparable with the mammalian pancreas, and containing both endocrine and exocrine cells, occurs in cyclostomes. Three separate associations of exocrine cells are distributed, respectively, along the hepatic blood vessels; as zymogen granule cells in the intestinal epithelium, and within the mesentery of *Myxine*.

THE RESPIRATORY SYSTEM

i In both cyclostomes and ostracoderms gaseous exchange involves gills in the hind part of a variously modified orobranchial region. In the pteraspidomorphs this extended back beyond the hind limits of the endocranium and opened on each side via a single branchial aperture at the level of the last gill. The Y-shaped impressions made by the visceral arches imply that these had anterior and posterior hemibranchs that were directly comparable with those of gnathostomes. In this respect, as in so many others, the pteraspidomorphs exhibit a structure from which that of all known vertebrates can reasonably have evolved. Previous suggestions that gills existed in front of what is the hyoid region of modern fishes are now discounted. The 'mandibular' arch supported the mouth and had no associated gill slit.

ii The branchial region of osteostracans was partially subdivided by interbranchial crests to give a series of branchial compartments, and the size of all the components decreased from the front backwards. However, as in lampreys, the situation was fundamentally different from that in gnathostomes since the gills were situated medially in relation to the visceral arches, whereas those in gnathostomes are situated laterally to these arches. Water flowing through the gill region passed first into epitrematic and hypotrematic chambers, above and below the gills respectively, and then to the outside via branchial apertures. The overall organization in anaspids was similar to this but the number of gill openings varied in different genera. From 8–15 external openings were distributed obliquely down the anterio-lateral region of the body.

iii Respiratory surfaces have to fulfil three general criteria in order to be effective. They must be thin, moist and highly vascular. In modern cyclostomes they comprise branchial lamellae situated within branchial pouches separated by interbranchial septa. Adult lampreys have seven pairs of these, each pair opening to the exterior by a pair of lateral branchial apertures. From 5–15 occur in the hagfish genus *Eptatretus* and again have independent openings to the outside. However, the 5–7 of the genus *Myxine* open via a single common ventral aperture (Fig. 2.11). In the ammocoete larva water is taken in at the mouth and is expelled through these pouches and apertures, but the pharyngeal region is transformed at metamorphosis and the branchial sacs then communicate with a special pharyngeal compartment. This additional respiratory feature of the adult is related to the fact that the animal's mouth is applied to the body of the prey, so that water is customarily both taken in and expelled through the distal apertures.

iv On either side of the gill pouch the gill lamellae are arranged in a semicircular manner to form a hemibranch or half-gill. Two hemibranchs, enclosing be-

Fig. 2.11 Diagram of the branchial region of *Myxine* showing the heart, ventral aorta, afferent blood vessels to the gill pouches and the single opening to the exterior.

tween them an interbranchial septum, form a gill or holobranch. Each hemibranch is separated from the interbranchial septum by a lymphatic space across which extend small, supportive, connective tissue strands.

v At the same time that respiratory movements are occurring in the branchial region, water is being forced independently into, and out of, the blind anterior nasopharyngeal pouch which has no counterpart in gnathostomes. The peculiar position of the single nasal aperture, and its unusual relationships with the nasopharyngeal pouch, reflect the shifting of these organs during embryonic development. This additional current maintains samples of the surrounding water in contact with the olfactory epithelium.

vi In *Myxine* the respiratory water current enters via the nostril, passes along the nasal duct to the olfactory organ, and then, via the nasopharyngeal duct, to the velar chamber at the anterior end of the long pharynx. It then passes, via afferent ducts, to the gills, and thence by efferent ducts to the external apertures. The velum acts

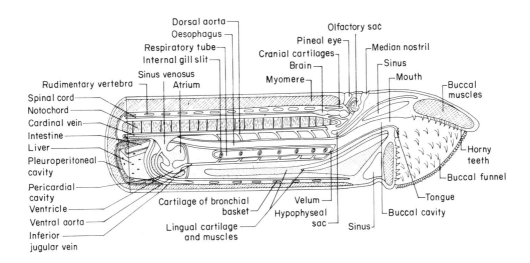

Fig. 2.12 Diagram of the anterior of a lamprey.

as a respiratory pump and has a crucial role in creating the respiratory current.

THE CARDIOVASCULAR SYSTEM

i Surprising though it may be, the work of Stensiö and his collaborators elucidated the blood vascular system of the ostracoderms so that it is as well known as that of the modern cyclostomes. In all essential features it resembles that of lampreys.

ii In lampreys a subintestinal blood vessel runs forward in the splanchnopleur beneath the intestine, breaks up into capillaries in the liver to form a hepatic portal system, and then hepatic veins proceed forwards to the heart. The most posterior chamber of the heart is a thin-walled sinus venosus. This leads to an auricle and this to a thick-walled ventricle. The overall length of the heart is longer than that of the pericardium within which it is contained so it is bent on itself to form an S shape. Both the entry to and exit from the ventricle are guarded by valves preventing blood from flowing backwards. From the ventricle the ventral aorta runs forward beneath the gill pouches to each of which it sends a pair of afferent branchial arteries. These pass to the gill arches where they breakdown into a capillary network. Paired efferent branchial vessels then gather up the oxygenated blood and take it to the dorsal aorta. This lies just below the notochord and distributes blood throughout the body. It is continued forwards into the head as the carotid arteries and behind the branchial region it sends

branches to the body wall and the pronephros (see below). The liver and intestinal tract are supplied by coeliac arteries; several mesenteric arteries occur more posteriorly; segmentally arranged vessels distribute blood to the more posterior kidney components and to the gonads. Posteriorly the aorta is continued into the postanal region as a caudal artery.

iii On each side of the dorsal aorta paired anterior and posterior cardinal veins lie in the body wall or somatopleur. These communicate with the sinus venosus of the heart by way of the Cuvierian ducts which cross the coelom in a transverse septum. This septum divides the coelom into an anterior pericardium and a posterior perivisceral splanchnocoel. Incomplete in the ammocoete larva, the division is complete in the adult lamprey.

iv The vascular system of hagfishes possesses various specialized features. The afferent branchial arteries carrying deoxygenated blood approach the gills from the side and form a circular canal around the efferent gill duct. This canal supplies a system of radial arteries to the primary gill lamellae where the blood vessels break down into a diffuse system of sinus-like cavities from whence the definitive respiratory exchange capillaries pass towards the central canal of each gill pouch (Fig. 2.13). The extremely thin walls of the exchange vessels are composed of connective tissue supported by special pilaster cells.

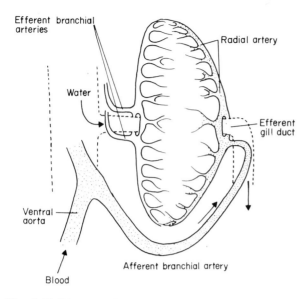

Fig. 2.13 Diagram of the vascular supply to a gill pouch in *Myxine*. (After Johnsen.)

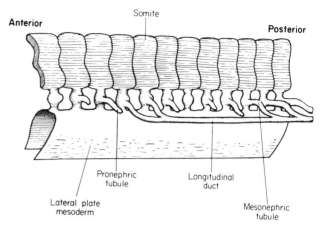

Fig. 2.14 Diagram showing the development of pronephric and mesonephric ducts.

The heart is aneural and in the presence of acetylcholine behaves like an embryonic heart. The left anterior cardinal vein is larger than the right and, in association with this, the right ductus Cuvieri is obliterated (the left had disappeared in fossil cephalaspidomorphs). Furthermore, apart from those in the skin and brain, capillaries are absent from the region in front of the auditory capsule where their place is taken by irregular blood sinuses.

THE URINOGENITAL SYSTEM

(a) The kidney

i The kidney may have arisen in freshwater provertebrates as a method of voiding excess water. That of cyclostomes, as in all vertebrates, forms from the residual mesoderm lying between the myotomes and the lateral plate musculature. Each pair of segmental units, or nephrotomes, typically contains cavities, the nephrocoel, which open into the splanchnocoel by ciliated funnels, the coelomostomes. During development the nephrocoels swell out into little cavities known as Bowman's capsules. Into each of these there projects a glomerulus formed by an arteriole derived from the dorsal aorta, and a venule leading to the cardinal vein.

The Bowman's capsule and glomerulus together make up a Malpighian or renal corpuscle.

ii From each capsule a tubule extends backwards and into the tubule of its next posterior neighbour. A collecting duct is thus formed on each side and, spreading back, meets its fellow in the midline and opens behind the anus on a small papilla (Fig. 2.14). The tubules arise in two sequential sets. An anterior series, opening into the pericardium and associated with the anterior cardinal vein, forms the pronephros or head kidney and its derivative pronephric duct extends back to the external papilla. This pronephros persisted in the adult cephalaspidomorphs, where it was protected by endoskeletal components, and is prominent in the myxinoid genus *Eptatretus*. It nearly disappears in adult lampreys and its function in adult myxinoids may not be excretory in the normal sense. In the later stages of all living vertebrates it is functionally replaced by a more posterior set of homologous tubules which form the mesonephros. These are associated with the posterior cardinal vein and as the component tubules open into the already existing pronephric duct this latter becomes the mesonephric duct. The mesonephros of *Myxine* is remarkable for the large size of its Malpighian corpuscles. These are restricted to the 30–35 intermediate segments and the segments lying both in front and behind this region lack corpuscles. It is possible that the long ureters have here taken over some of the functions of the renal tubules in other vertebrates. In *Eptatretus* there is a more or less complete series of pronephric and mesonephric derivatives.

(b) The gonads

The reproductive systems of cyclostomes are amongst the simplest of all vertebrates. In very young animals the rudimentary testes or ovaries are similar. They arise from paired rudiments but the adult gonad is a single unpaired structure suspended by a mesentery. This last is known as the mesorchium in males, and the meso-varium in females. The gonads never unite with renal derivatives to form compound structures such as a vas deferens, nor is there an oviduct. The sperm or ova are simply shed into the coelomic cavity whence they pass to the outside via a genital pore in myxinoids. In the male lamprey the approach of sexual maturity is accompanied by a narrowing of the urinogenital sinus which opens by a small pore at the end of a tube forming a 'penis'. In contrast, that of the female forms a large vestibule with a 'vulva' and wide opening.

THE ENDOCRINE ORGANS

(a) The hypophysis

In all vertebrates the hypophysis (= pituitary gland) has dual origins. In the adult, a neural derivative, the neurohypophysis, is closely apposed to an ectodermal derivative from the stomodaeum, the adenohypophysis. Ostracoderms had a well-marked hypophyseal fossa suggesting that the organ was certainly present in the earliest vertebrates. That of lampreys lies between the floor of the diencephalic region of the brain and the nasopharyngeal pouch. The adenohypophysis is situated below the neurohypophysis and its cells are arranged in cords. The pituitary of *Myxine* is unusual and perhaps degenerate. Similarities between the β cells of the adenohypophysis and the epithelial spider cells of the nasopharyngeal pouch have led to the suggestion that they are homologous.

(b) The thyroid gland

As noted above (page 7), in the ammocoete larva a ciliated groove runs along the floor of the pharynx and receives the mucus secreted from an endostyle. This is a hollow, ventral downgrowth from the pharynx containing four rows of glandular cells. During metamorphosis important changes occur. The endostyle closes up, its ciliated cells disappear, and it gives rise to the thyroid gland in the adult. This comprises scattered follicles located around the ventral aorta rather than the en-

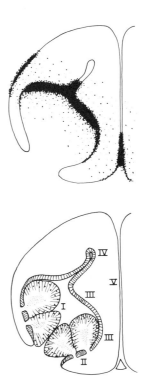

Fig. 2.15 The distribution of five cell types in the ammocoete endostyle and (above) the distribution of iodine as shown by autoradiography. (After Gorbman & Creaser.)

capsulated structure of amniotes. Both the endostyle and the adult thyroid gland are characterized by their affinity for iodine and the adult cells produce a thyroxine-like hormone. Five cell types are topographically differentiated in the larval endostyle and autoradiography reveals that iodine accumulates in Type III and, to a lesser extent, in Type V (cf. Fig. 2.15). It is thought that the adult thyroid follicles develop from types II and IV.

(c) The parathyroid glands

During the embryonic phases of all vertebrates small budlike masses of cells proliferate from the dorsal and ventral regions of the pharyngeal pouches. In tetrapods certain of these give rise to the parathyroid glands (see pages 37 and 116). In cyclostomes their small size and diffuse nature, together with the difficulty of extirpating them completely, has rendered it difficult to determine whether they are parathyroids.

(d) The adrenal tissues

i Unified adrenal glands comparable with the compound structures of mammals do not occur in cyclostomes. Instead there are two clearly distinguished series of structures. The adrenocortical or inter-renal material consists of small, irregular, lobelike aggregations distributed along the posterior cardinal veins, the renal arteries and other arteries close to the mesonephros. They appear to arise from coelomic epithelium at the hind end of the pronephros in much the same way as in other vertebrates. As elsewhere, they are responsive to adenohypophyseal hormones. It has been suggested that adrenocortical structures are derived from coelomostome derivatives which retain, in the adult, crucial ion controlling functions.

ii The other, chromaffin, series, homologous with the adrenal medulla of mammals, extends from near the anterior part of the branchial region, opposite the second gill cleft, to the tail. It takes the form of small strips of tissue along the course of the dorsal aorta and its branches.

(e) The thymus gland

Stensiö interpreted certain structures in the branchial region of fossil cephalaspidomorphs as thymus derivatives. A transitory thymus occurs in the ammocoete but no thymus gland has been observed in myxinoids and it has not proved possible to stimulate them to respond immunologically.

(f) The endocrine pancreas

As already emphasized, a unified pancreas is absent from cyclostomes. Instead, in lampreys a few small masses of endocrine cells lie buried in both the liver and the intestinal wall. In hagfishes an 'islet' organ forms a thickening on the distal end of the bile duct. Each islet is separated from its neighbours by connective tissue and the whole organ is permeated by fine capillaries.

THE SENSE ORGANS

(a) The olfactory organ

i Olfaction provides information about a wide variety of molecules within the surrounding medium. It is therefore a very important sensory modality in aquatic animals. The olfactory apparatus of the lamprey is a

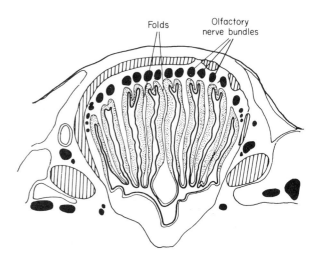

Fig. 2.16 Transverse section of the olfactory apparatus in *Myxine*.

unique structure. The single dorsal nostril leads into a canal which connects posteriorly with a small olfactory sac lying just in front of the brain. The single median olfactory mucous membrane is, however, associated with a pair of olfactory nerves and this, together with its symmetrical disposition in the midline, suggests that the olfactory sac has been derived from paired structures during phylogeny.

ii A diagram of the apparatus in *Myxine* is shown in Fig. 2.16. Adjacent layers of olfactory epithelium are separated from each other by a space which is in direct communication with the nasopharyngeal duct. The sensory cells within the epithelium give rise to nerve fibres that pass to the dorsal surface, collect into bundles and then penetrate the neurocranium through separate foramina. Those from each side enter the ipsilateral olfactory lobe of the brain. In Amphioxus all the sensory fibres within the dorsal roots of the central nervous system originate from sensory cells, but in vertebrates it is only the olfactory nerves which arise in this way.

(b) Photosensitive structures

PHOTORECEPTORS IN THE SKIN
Cyclostomes have light-sensitive cells in the skin as well as in the eyes. In *Myxine* exploration of the body surface with a light source demonstrated that the light sense is distributed over the entire body but that the end organs are generally sparse and only numerous near the front of

the head and in a cloacal-caudal region. Transection of the spinal cord produces different results in *Myxine* and *Lampetra*. In *Myxine* the body in front of the transection no longer exhibits responses when light is shone onto the tail of the animal. In contrast the region anterior to the transection still responds in *Lampetra* and the animal swims away. This suggests that the light sense is here mediated by the lateral line nerves not the spinal cord.

THE PINEAL AND PARAPINEAL EYES

Most vertebrates possess a structure derived from the diencephalic roof that is variously a pineal eye or pineal organ. The presence of a pineal foramen in Osteostraci demonstrates its presence in ostracoderms. Amongst cyclostomes the pineal is absent from *Myxine* and it takes the form of a pineal photoreceptor overlying a second, smaller, parapineal structure in lampreys (Fig. 2.17). These are generally considered to be the right- and left-hand components of a pair. The component photoreceptor cells of the pineal bear a strong morphological resemblance to retinal rods and cones and considerable evidence now shows that they participate in the control of motility and adaptive pigmentation. It is not yet clear whether the indole melatonin, characteristic of the pineal in birds and mammals, is involved.

THE PAIRED EYES

i The embryological development of the definitive eyes

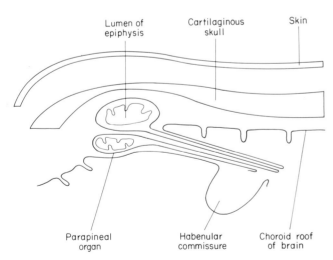

Fig. 2.17 Diagrammatic sagittal section showing the relationship of the pineal (epiphysis) and the parapineal organs. (After Wurtman *et al.*)

of vertebrates parallels that of the pineal and parapineal derivatives. The retina with its sensory cells develops as an evagination from the diencephalic region of the brain and the optic nerve or tract is, therefore, unlike other peripheral nerves and, together with the retina, represents a portion of the brain.

ii Paired orbits demonstrate that the ostracoderms had eyes. Those of myxinoids have no lens, iris or eye muscles and the retina is both thin and unpigmented. Although it has three or four recognizable cell types these are not very distinctly stratified and on morphological grounds alone it might be concluded that the eyes are functionless. Indeed, the very existence of nerve fibres within the optic tract has been questioned. The eye of the ammocoete larva is covered by skin and is only transformed into a functional eye during metamorphosis.

iii In adult petromyzontids, the eyes, although somewhat primitive, exhibit the principal features seen in all other vertebrates (Fig. 2.18) and their lateral position gives each a visual field encompassing some 170°. They are broadly spherical in appearance, although the external segment of the sphere is somewhat flattened, and surrounded by prominent blood sinuses which penetrate to the middle of the iris. Intra-ocular blood vessels are absent.

The cornea consists of two distinct layers separated by a gelatinous substance and the innermost of these is continuous with the fibrous sclerotic capsule that surrounds the entire eyeball. The method of accommodation is unique. Two anterior myotomes contribute to a cornealis muscle which pulls on the cornea, flattens it, and moves the lens nearer to the retina. The choroid coat is thin in front but thickened towards the hind pole of the eye where it contains numerous arterioles. In the equatorial region of the eyeball it is separated from the sclerotic by a tissue containing vascular lacunae, melanophores and large vesicular cells. The retina is about 200 μm thick and contains both rod and cone cells. There are bipolar, amacrine and rather few ganglion cells.

(c) The equilibratory organs

The otic region (ear) of cyclostomes is generally considered to have regressed during phylogeny—that of hagfishes more so than those of lampreys. Only the membranous labyrinth is present and neither utricular nor saccular sensory regions are distinct. The presence

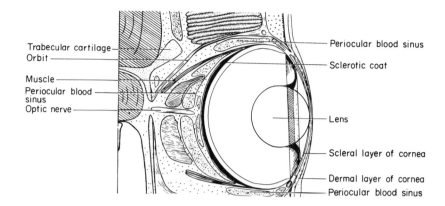

Fig. 2.18 Horizontal section through the eye of *Lampetra fluviatilis*. (After Franz.)

Labels on figure: Trabecular cartilage, Orbit, Muscle, Periocular blood sinus, Optic nerve, Periocular blood sinus, Sclerotic coat, Lens, Scleral layer of cornea, Dermal layer of cornea, Periocular blood sinus

of only two vertical semicircular canals in petromyzontids, each with an ampulla at the lower end, is clearly an ancient feature as a third canal is absent from the ear of fossil cephalaspidomorphs. *Myxine* has a single canal in the form of a thick ring (Fig. 2.19), which has been interpreted as the posterior vertical canal, together with a spacious chamber corresponding to the utriculus-sacculus complex. However, the presence of

two cristae, and two slightly swollen regions, the 'ampullae', at opposite ends also suggests that it is the fused remnant of the two petromyzontid canals. Sensory cells in the basal region form a so-called macula acoustica. A tiny endolymphatic duct completes the list of recognizable components.

(d) The tentacles

The tentacles are almost the only noteworthy external processes on hagfishes. There are two oral and four nasal ones in *Myxine*; two oral and two nasal ones in *Eptatretus*. They bear both large and small sense cells that are sensitive to tactile stimuli.

(e) Lateral line organs

All primarily aquatic vertebrates have special mechanoreceptors in the skin known as lateral line organs (Figs 2.20 and 2.21). It is clear that such structures also

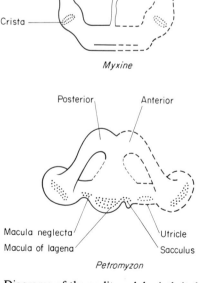

Fig. 2.19 Diagrams of the auditory labyrinth in hagfish and lampreys. (After de Burlet.)

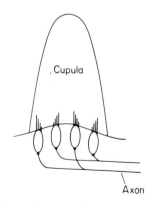

Fig. 2.20 Diagram of a lateral line neuromast.

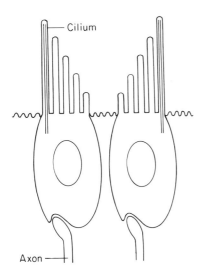

Fig. 2.21 Diagram showing the organization of two hair cells in a neuromast organg.

occurred in ostracoderms. In cyclostomes, many bony fishes and amphibians, they comprise individual organs that are not joined by canals. Each single epidermal organ, called a sensory hillock or neuromast, consists of a cluster of pear-shaped sensory cells surrounded by long, slender, supporting cells. Sensory hairs at the apex of the sensory cells project into a jelly-like substance, the cupula, which is secreted by the neuromast. Minute displacements of the cupula parallel to its base (shearing) displace the sensory hairs and induce receptor potentials. There are, in fact, two types of sensory unit in each neuromast, each activated by water currents that displace the cupula in a particular direction. Currents striking the cupula perpendicular to these axes of maximal response are ineffective. The structures enable both the detection and location of moving animals by virtue of the water disturbances which they create.

THE 'ELECTRIC ORGANS'

The head shield of osteostracans bore areas of tissue whose function remains controversial. They were supplied by one dorsal and six lateral nerves which were exceedingly thick. On the one hand it has been suggested that the structures were electric organs providing a defence against predators such as the eurypterid chelicerates. On the other hand it has been suggested that they were components of the lateral line system.

THE CENTRAL NERVOUS SYSTEM

(a) The spinal cord

i The central nervous system of cyclostomes again exhibits the basic segmentally repetitive plan that is common to all vertebrates. In ammocoetes, few, if any, capillaries actually penetrate the substance of the central nervous system and its ribbon-like appearance has been attributed to the pial capillary network being the principal source of metabolic substrates. Special nerve cells lying just outside the neural canal send one fibre to a peripheral sensory cell and another into the central nervous system. These nerve cells lie on the track of the sensory nerves and form swellings or ganglia—the dorsal root ganglia. Within the spinal cord the incoming sensory fibres branch in a T or Y fashion and thereby give rise to ascending or descending branches within the dorsal funiculus. In gnathostomes the homologous branches give off collaterals but such collaterals are rare in cyclostomes.

ii Unlike the situation in all other vertebrates, the ventral roots remain separate from the dorsal ones. They are composed of nerve fibres whose cells of origin lie in the spinal cord but the peripheral distribution of the fibres is not purely segmental. In *Myxine* some enter the myoseptum immediately adjacent to the ventral root and innervate muscle fibres in both of the myotomes that are attached to it. Other fibres pass to the next anterior myotome, innervate its muscle fibres, and then pass into the next one, or perhaps two, myotomes. Some fibres pass to the mucous glands.

Three types of cell contribute axons to the ventral root (Fig. 2.22). The most abundant lie lateral to the ventral root fibres in the spinal white matter—often at the lateral border of the grey matter. The majority are

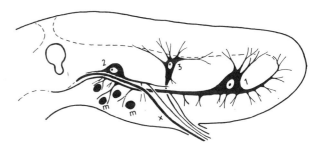

Fig. 2.22 Types of cell within the spinal cord. 1 and 2, somatomotor; 3, visceromotor; m, Muller type axons. (After Bone.)

25 μm in size and probably innervate the large, fast muscle fibres of the myotome. The second type are smaller, less abundant, and lie medially to the point of exit of the root. These are thought to innervate the slow muscle fibres. The third type comprise multipolar elements lying dorsally within the spinal grey which are probably the cells of origin of visceral fibres.

The majority of commissural connections between the two sides of the spinal medulla occur in the ventral commissure. Most of the somatic neurons send dendritic branches across in this commissure and these then divide into ascending and descending branches. There are, however, some decussations in a dorsal commissure. The most conspicuous are dendritic processes of internuncial neurons.

iii Although there are no sympathetic ganglionic trunks the ventral spinal nerves give off visceral branches which are comparable in diameter to the other ventral root fibres. They leave the ventral root at the level of the notochord, pass first of all in a medio-caudal direction, and then ventrolaterally to reach the viscera. Pharmacological investigations show that, with the exception of the hagfish heart, the viscera of cyclostomes respond to acetylcholine and adrenalin in a way that is analogous to the autonomically innervated structures of gnathostomes.

iv Large coordinating neurons known as Muller and Mauthner cells occur in both the spinal cord and the brainstem. The axons of the Muller cells pass up and down on the side of the nervous system in which their cell bodies are situated—they are ipsilateral projections. Those of the Mauthner cells decussate—cross to the other side.

(b) The brain

i There appear to be major differences between the brains of pteraspidomorphs and those of all other agnathans. This conforms to the general thesis that the pteraspidomorphs are the most primitive known vertebrates. Impressions of the brain are preserved in some fossil specimens and indicate that there were paired nasal capsules and that the brain was not flexed in either a dorsal or ventral direction. The overall appearance is, therefore, simpler than that in any other known vertebrate and it was little more than a simple nerve cord swollen in certain regions. The hindbrain was a typical medulla bearing an anterior cerebellum. The midbrain was primitive with no evidence of optic lobes.

ii Reconstructions of the cephalic region of fossil cephalaspidomorphs suggest that the overall structure of the brain was not dissimilar to that in living petromyzontids. Although all exhibit a basic plan homologous with that in gnathostomes they have many unique features.

THE MEDULLA OBLONGATA
In all vertebrates the medulla oblongata contains cell aggregations, nuclei, from which the cranial nerves originate and at which they terminate. It also contains both ascending and descending fibres, and the more or less diffuse components of the brainstem reticular formation. These last occupy a rather irregular volume within the framework provided by the principal fibre tracts and the cranial nerve nuclei.

In general terms the sensory and motor units are organized in a definite dorso-ventral array (cf. Fig. 3.29). Relatively prominent dorsolateral sensory aggregations are associated with incoming fibres from the lateral line system in lampreys but these are less prominent in hagfishes. The general somatic afferent systems are represented by scattered cells. No viscero-sensory zone is delimited as a distinct entity in lampreys, but the ventrally situated motor units do comprise medial somatomotor and more lateral visceromotor aggregations. The somatomotor zones of each side are separated from each other by an inferior median sulcus.

THE CEREBELLUM
The gross appearance of the cerebellum in the different vertebrate classes varies considerably in association with body size, and the relative precision of the animal's motor ability. In *Petromyzon* the corpus cerebelli, represented by a compact mass of cells confluent behind with the periventricular grey of the medulla, is entirely buried and forms the deep part of the dorsolateral wall of the fourth ventricle. Those of hagfishes are even more rudimentary.

THE MESENCEPHALON
The midbrain of the ammocoete comprises two dorsal and two ventral zones separated by a sulcus limitans. It contains the brain nuclei of the oculomotor and trochlear nerves, cranial nerves III and IV. Dorsally there are the components of an optic relay region, the optic tectum, which is not as prominent as those of most gnathostomes, and a semicircular torus which is a nuclear aggregation related to the vestibulo-lateral system.

THE DIENCEPHALON

In all vertebrates the diencephalon consists of those structures which surround the third ventricle and are derived during ontogeny from the hind part of the prosencephalon. As such they probably represent a forward extension of sensori-correlative centres beyond the anterior limits of the ancestral, metamerically segmented, myotomal region. They therefore lack those imprints of branchiomerism which are detectable, in the form of cranial nerve nuclei, within more posterior regions. The sulcus limitans, which provides a topographical boundary between the embryonic alar and basal plaques from which sensory and motor components are respectively derived, curves downwards after entering the diencephalon and is widely considered to end at the preoptic recess. During ontogeny the diencephalic walls give rise to prominent nuclear aggregations—the thalami. The pineal eye is situated dorsally in association with a habenular region; the hypophysis is closely apposed to the ventral hypothalamic region.

Within the thalamus of *Petromyzon* there are two pairs of geniculate nuclei associated respectively with the optic and vestibulolateral systems. The overlying habenular complex is markedly asymmetrical, perhaps reflecting the differential development of pineal and parapineal eyes, and the underlying preoptic nucleus of the hypothalamus sends secretory perikarya to end within the neurohypophysis which is apposed to the pars intermedia of the adenohypophysis.

THE TELENCEPHALON

Although the cyclostome telencephalon theoretically provides an analogy with that of ancestral vertebrates, it actually differs from that of all other extant vertebrates in its foreshortened and high appearance. This is particularly true of *Myxine* and reflects the unusual morphogenetic movements of the buccal region. The

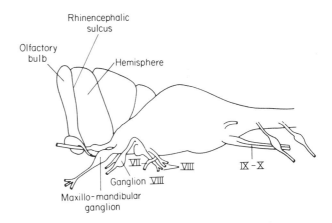

Fig. 2.23 Lateral view of the brain of *Myxine*. (After Jansen.)

olfactory bulbs are well developed, closely apposed to the olfactory organ, and delimited from the hemisphere by a transverse sulcus similar to that separating the diencephalon from the telencephalon. The hemispheres have an evaginated origin during ontogeny and are, therefore, comparable with those of Amphibia and Choanichthyes and quite distinct from those of bony fishes. The component structures remain controversial but, in broad terms, there is a medial hippocampal region, a dorsal pallial and a lateral pyriform region, which can be compared with those of other classes (Fig. 2.23).

FURTHER READING

HARDISTY M.W. (1979) *The biology of cyclostomes.* Chapman and Hall, London.
HOAR W.S. & RANDALL D.J. (eds) (1969) *Fish physiology,* 7 vols. Academic Press, London.

3 · Class Chondrichthyes

This, and all the remaining classes of vertebrates, possess jaws and are referred to collectively as gnathostomes.

Synopsis
Subclass Elasmobranchii
  *Order Cladoselachii   —*Cladoselache*
  Order Selachii   —*Squalus*
  Order Batoidea   —*Raia*
Subclass Holocephali
  Order Chimaeriformes —*Chimaera*

INTRODUCTION

i The world fish fauna includes two rather distinct taxa whose members have wholly cartilaginous skeletons. On the one hand there are the rather rare, oceanic, rabbit fishes such as *Callorhynchus* and *Chimaera*. These are members of the Holocephali (Fig. 3.1). On the other hand, there are the more diverse and numerous elasmobranchs varying in appearance from the dogfish, *Scyliorhinus*, or shark *Carcharodon*, to the sawfish *Pristiophorus*, or skates and rays such as *Raia*, *Rhinobatis* and *Trygon* (Fig. 3.2). Many sharks are large, and the whale shark, *Rhincodon typus*, is, at 17 metres in length and 40 tons in weight, the largest known fish. In contrast, *Euprotomicrus* has an adult length of only 20 cm. The appearance and gross morphology of a particular species of fish, irrespective of its particular taxonomic position, reflects precise adaptations to both its method of locomotion and way of life. Consequently, there are numerous examples of convergent evolution, our knowledge of the interrelationships of extant taxa has been the subject of much controversy, and the early evolutionary history of all groups remains contentious.

ii Following Watson's classic paper, the branchial region of the fossil acanthodians (Fig. 3.3) was considered to include a functional hyoidean gill slit. They were, therefore, thought to represent an evolutionary stage that was intermediate between the agnathans and the true fishes, and were ascribed to a separate aphetohyoidean grade of development—a term meaning free hyoid. Nevertheless, the hyoidean gill slit always appeared to be greatly constricted by the size of the mandibular arch and it was difficult to see how it could have functioned efficiently. Furthermore, Stensiö pointed out that the lateral line system seemed to traverse what was described as an open gill slit and Miles demonstrated that the dorsal part of the hyoid arch, the hyomandibular, articulated with the otic capsule and must have been involved with the jaw suspension. These observations elicited much discussion about the taxonomic relationships. Watson necessarily considered them to be unrelated to any modern taxa, Miles thought that they were related to bony fishes, whilst Stensiö, Holmgren and Jarvik considered them to be related to cartilaginous fishes.

iii Another group of armoured fishes appear to have

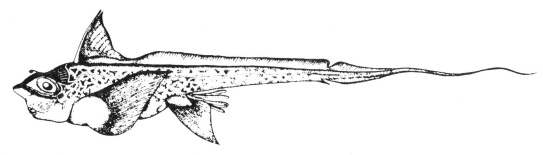

Fig. 3.1 Lateral view of a *Chimaera*.

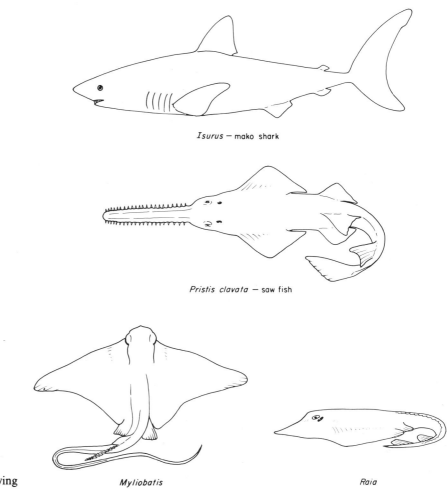

Isurus — mako shark

Pristis clavata — saw fish

Fig. 3.2 Some elasmobranchs showing the diversity of form.

Myliobatis

Raia

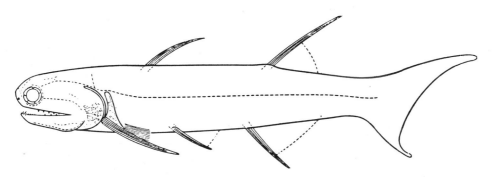

Fig. 3.3 *Ischnacanthus*, a Devonian acanthodian. (After Watson.)

Fig. 3.4 *Coccosteus*, a Devonian arthrodire.

been successful carnivores during the Devonian period. These had a prominent neck joint in their armour and were called arthrodires (Fig. 3.4). Some possessed a covering of small polygonal plates or tessarae but later genera had a well-defined system of large plates. Many genera had an overall appearance that is strongly reminiscent of modern skates and rays and, in the absence of any definitive criteria which would determine whether or not one was dealing with convergent evolution, they are quite widely described as elasmobranchiomorphs. One particular genus, *Ctenurella*, was then shown to strongly resemble modern chimaeras. These holocephalans possess a number of characters that had formerly been considered unique, such as paired rostral cartilages and the peculiar forms of both the copulatory organs and dorsal fins. The occurrence of apparently comparable characteristics in *Ctenurella*, therefore led to the suggestion that the chimaeras are surviving arthrodires and also that sep-

arate lineages, respectively ancestral to cartilaginous and bony taxa, date from the earliest known fossil fishes. The existence of so many examples of convergent evolution renders any final assessment of this suggestion difficult.

THE INTEGUMENT

i The majority of selachians are active carnivores at the apex of oceanic food chains and have few predators. Consequently they lack extensive external armaments. Nevertheless, the skin is tough and formed from inert protein, whilst characteristic placoid scales or denticles are scattered over the entire surface (Fig. 3.5). These are homologous with teeth and composed of dentine laid down by a layer of odontoblasts situated around a pulp cavity. The outside of the dentine is covered by a layer of enamel secreted by the overlying ectoderm but once the denticles pierce this ectoderm no further enamel is deposited. The characteristic microscopic appearance of the dentine reflects its penetration by fine processes from the odontoblasts.

ii An examination of the scales of the earliest sharks suggests that these were complex and that modern placoid scales are derivative structures which only appear in the fossil record at a much later date. Stensiö

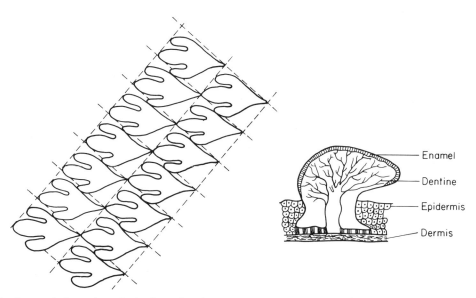

Fig. 3.5 The lattice-work formed on the body surface by neighbouring scales and a single denticle shown diagrammatically in longitudinal section. (After Marshall.)

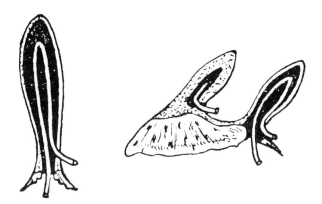

Fig. 3.6 A single lepidomorium from which scales may have arisen by fusion and an early scale formed by fusion of two lepidomoria. (After Stensiö.)

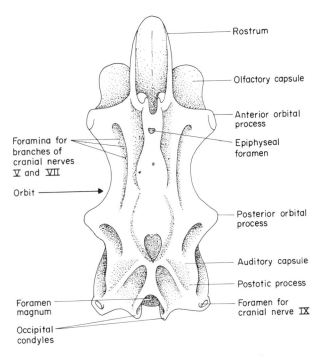

Fig. 3.7 Dorsal view of the skull of *Squalus acanthias*.

and Orvig presented evidence which suggests that they resulted, during phylogeny, from the progressive fusion of small elements to give larger ones (Fig. 3.6). They envisage that the skin was originally covered with minute structures, lepidomoria, each containing a single vascular loop and a simple pulp cavity. They were frequently fused at their bases but remained separate at their crowns. The placoid scale is envisaged as a later development. The only remaining evidence for its phylogenetic history lies in the complex ramifications of its pulp cavity and the series of ridges or cusps which occur on the outer surface. Detailed schemes suggest that the placoid scales are, indeed, more advanced phylogenetic features than the cyclomorial scales of bony fish. As the conical teeth of early vertebrates are universally considered to be derived from placoid scales, these too had a comparable phylogenetic history.

SKELETAL SYSTEM

(a) The cranial region

SELACHIANS

i The skeleton of Chondrichthyes is, of course, entirely cartilaginous. For descriptive purposes the chondro-cranium of selachians can be differentiated into an anterior ethmoid region with the nasal capsules separated by a median nasal septum; an orbito-temporal region; an otic region containing the auditory capsules, and a posterior occipital region that is connected to the vertebral column and surrounds the foramen magnum (Fig. 3.7). The brain extends from the nasal septum back through the three more posterior regions and is pro-

tected by them. The main part of the cranium develops from sclerotome, as do the vertebrae, but the ventral components are derived from neural crest material and represent the dorsal elements of premandibular and mandibular visceral arches (q.v.).

ii As in cyclostomes, the first chondrifications to appear during embryological development are the paired parachordals. These are situated on either side of the notochord (Fig. 3.8), and in the vicinity of the otic capsules their lateral edges extend further sideways than elsewhere, forming the basiotic laminae. The base of the skull in the more anterior, prechordal region appears as paired trabeculae which are the remnants of visceral arches in the premandibular segment. The nasal septum is formed by the union and upgrowth of the anterior ends of the trabeculae in association with the cartilages of the nasal capsules. The side walls in the orbito-temporal region arise from paired orbital cartilages and the otic capsules generally arise as independent cartilaginous plates but soon fuse with the parachordals. As the trabeculae fuse in front, and with the parachordals or their derivative basal plate behind, a median hypophysial space or basicranial fenestra is delineated. Through this the hypophysis protrudes and the internal

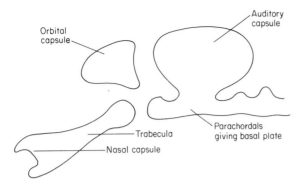

Fig. 3.8 Diagrammatic lateral view of a developing chondrocranium showing the contribution of trabeculae and parachordals. (After Goodrich.)

carotids pass. In the selachians, and also in bony fishes such as *Amia*, *Acipenser*, *Polypterus* and the Dipnoi, the trabeculae remain widely separated in the orbito-temporal region giving a so-called platybasic skull in which the brain extends forward to the olfactory capsules. In front of the jaws, additional labial cartilages sometimes form at the side of the head.

HOLOCEPHALIANS

i The general development of the rather specialized chondrocranium in chimaeras (Fig. 3.9c) parallels that of selachians. Thus trabeculae and parachordals form trabecular and basal plates and the trabecular plate is perforated by a small hypophysial foramen. There are no internal carotid foramina as these arteries fail to enter the cranial cavity. In selachians the anterior region of the skull is not chondrified and the epiphysis lies below an extensive fontanelle. In holocephalians this is closed by ethmoidal canals. During development these appear as longitudinal, horizontal tunnels entirely enclosed by cartilage. Situated immediately above the roof of the nasal capsules, they open on to the dorsal surface of the snout in front and into an interorbital space continuous with the orbits behind.

ii The nasal capsules lack cartilaginous hind walls and are, therefore, continuous with the cranial cavity. The occipital region is of interest because there is a special craniovertebral joint. The posterior edge of the basal plate forms a saddle-shaped surface that articulates with the centrum (q.v.) of the first vertebra. At the same time the posterior surface of the chondrocranium bears a facet which articulates with the neural arch of the first vertebra. This point of special flexibility between the skull and vertebral column may be related to the fact

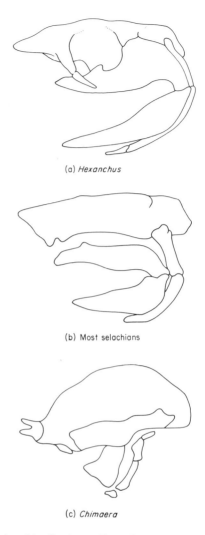

(a) *Hexanchus*

(b) Most selachians

(c) *Chimaera*

Fig. 3.9 Amphistylic, hyostylic and autostylic jaw suspensions in *Hexanchus*, many selachians and *Chimaera* respectively.

that, as in batoid Selachii, the anterior vertebrae are immovably fused together.

(b) Jaws and branchial arches

SELACHIANS

i In selachians the gill slits open separately from the pharynx to the exterior and the first one on each side is reduced to form the spiracle. Within the septa between each successive pair of gill slits a cartilaginous branchial arch develops. The first of these between the spiracle and the mouth, becomes the jaws. Those behind are composed of several cartilages. The first, the hyoid, is attached to the cranium in the auditory region, the remainder lie alongside the occiput and the vertebral column. They are of neural crest origin, lie deep in the body wall, and are not directly homologous with the compound branchial basket of cyclostomes.

ii The component cartilages comprise dorsal pharyngobranchials, then epibranchials, ceratobranchials and ventral hypobranchials which are joined together in the midline by unpaired basibranchials. The hyoid arch has the usual ventral elements but only one dorsal element, the hyomandibular, which corresponds to the epibranchials. Each arch apart from the last bears a single row of cartilaginous branchial rays on the epi- and ceratobranchials. These radiate outwards in the gill septum between successive gill slits. In many selachians, curved extrabranchials lie outside them.

iii In the early sharks, and in a few modern genera such as *Cestracion*, *Hexanchus* and *Heptanchus*, the upper jaw, or palato-pterygo-quadrate cartilage, articulates both on the cranium and on the most dorsal cartilage of the hyoid arch—the hyomandibular. Up to four jaw processes articulate on the cranium, three in the otic region and one, the ethmoid process, in the orbital region. The hyomandibular articulates with the otic capsule and the jaw (Fig. 3.9a). In all other extant sharks the articulation with the cranium is lost and the jaw is suspended by the hyoid arch and by ligaments and muscles attaching it to the cranium (Fig. 3.9b). These respectively comprise the amphistylic and hyostylic methods of jaw suspension.

HOLOCEPHALIANS
The situation in chimaeras differs from the foregoing in a number of respects. They lack a spiracle and the definitive gill slits are covered by a flap of skin, the operculum, that is supported by branchial rays on the hyoid arch. They also have a specialized method of jaw suspension which has been envisaged as a development associated with the presence of permanent grinding tooth plates adapted for consuming hard food. The jaws are short, strong and fused in front, while the palato-pterygo-quadrate is fused to the ethmoid and orbital regions of the cranium in front and to the auditory capsule behind. Since the hyoid arch remains free and is not involved in the suspension of the jaws, this suspension is termed autostylic, or, to emphasize that it

probably evolved independently of that in Dipnoi and tetrapods, holostylic.

(c) The axial skeleton

SELACHII
i The structure of the selachian vertebral column is again most easily described in terms of its embryological development. From an early stage the notochord of all vertebrates is surrounded by a fibrous sheath confined within an outer elastic membrane and separated from the notochord by a less distinct inner elastic membrane. In selachians mesoblastic cells from the skeletogenous layer outside the external membrane invade the sheath and grow, surrounding the notochord. The sclerotomes yielding this mesoblastic tissue, from which the vertebral column subsequently develops, are, initially, metamerically arranged (Fig. 3.10). Small intersegmental blood vessels arising from the dorsal aorta and cardinal veins extend upwards between successive sclerotomes. During development, cartilage forms in the sclerotomes and gives rise to the components of the vertebrae. They include a centrum, surrounding and containing the notochord, and arcualia, four paired blocks of cartilage lying outside the external elastic membrane near to the upper and lower surfaces of the notochord.

ii The centrum may develop as a cartilaginous ring in continuity with the arcualia, or cartilage may spread from their bases. The definitive centra alternate with the original segments and hence with the

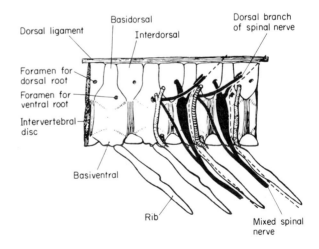

Fig. 3.10 Left side of the vertebral column in *Squalus acanthias*. (After Goodrich.)

myotomes. Between them, where they meet in the middle of the original segments, the intervertebral ligament is formed.

iii The pair of dorsal arcualia that form within the posterior half of each segment are known as basidorsals. They grow up on either side of the nerve cord, fuse with each other, or with independent supra-dorsal cartilages above the nerve cord, and form the neural arch. The homologous ventral pair are the basiventrals. They grow ventrally and, in the tail, surround the caudal artery and vein and fuse to form the haemal arches. This does not happen in the trunk region. The dorsal and ventral arcualia which form in the anterior half of each original segment, the interdorsals and interventrals, are always smaller than the basalia. However, various differences occur in different regions of the vertebral column in different species. In particular, in some parts of the column, notably near the end of the tail, two centra may form in association with a single myotome, one to each set of arcualia, giving a condition known as *diplospondyly*.

HOLOCEPHALI

The Holocephali retain a persistent and unconstricted notochord. The thick fibrous sheath of *Chimaera* is invaded by mesoblastic cells through the ruptured elastica externa. These then gather into a middle zone within the sheath and calcified rings develop which are more numerous than the original segments (Fig. 3.11). Such rings are absent from *Callorhynchus*, but very strong and closely packed in the Mesozoic *Squaloraia*. Their reduced condition in, or absence from, deep sea forms has been attributed to the use of paired fins, rather than the tail, for swimming.

Fig. 3.11 Vertebral rings of *Squaloraia*—a holocephalian.

(d) Pterygiophores

Selachian fins, including the caudal one, develop from metameric derivatives in the embryo. However, amongst extant shark-like genera the adult structures appear to be more restricted than their myotomes of origin. A dorsal fin derived during ontogeny from, perhaps, twelve segments, may only appear to extend over six myotomes. Each segment of fin consists of two or more rods of cartilage, the pterygiophores, that are connected end-to-end. The distal ones lie between very thin collagenous rods, the ceratotrichia, which extend to the edge of the fin and stiffen it. Although many develop in each segment, there is only one corresponding radial muscle which is attached by a broad tendon to the ceratotrichia and which, when it contracts, bends the fin towards its side (Fig. 3.12).

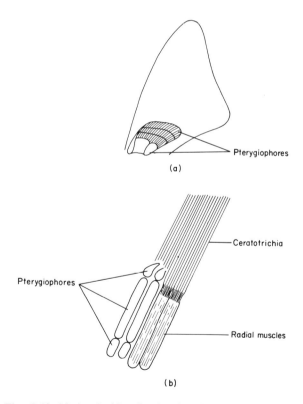

Fig. 3.12 (a) A selachian fin showing the organization of pterygiophores, and (b) the organization of pterygiophores and ceratotrichia.

MUSCULAR SYSTEM

(a) General considerations

It has already been emphasized that during both phylogeny and ontogeny somatic musculature arises from myotomes, whereas the visceral musculature is derived from lateral plate mesoderm. The myomeres are the main agents of locomotion and have differentiated into several longitudinal bundles. The dorsal bundles constitute the epaxial muscles and the ventral ones hypaxial muscles. Internally the myosepta between the myotomes are attached to the vertebral column and the fibres of a myomere extend from one myoseptum to the next. The actual arrangement is rather complicated. The myomeres are drawn out into long forward- and backward-directed cones—the point of one fitting into the hollow of the next. Stout tendons project from the myosepta to the apices of these cones. The white fibres, which form some 82% of the total in *Scyliorhinus*, converge on these tendons in a complex pattern which results in all of them shortening by about the same fraction, and at the same rate, when the fish bends.

The eagle and manta rays (Myliobatidae and Mobulidae) use their pectoral fins more like wings but typically the myotomes on each side of the body act antagonistically, those on one side contracting as those on the other side relax. Their contraction shortens the body wall. Many other muscles also act in antagonistic groups and are named according to their function. Those raising a structure are levators, those depressing it depressors. If they move it towards the midline they are adductors, if away from it, abductors. If they flex a structure they are flexors, if they extend it they are extensors.

(b) Cranial myotomes

Most of the cranial musculature, including that of the jaws and branchial arches, is derived from lateral plate mesoderm. As in ammocoetes the somites of the three pro-otic segments form the extrinsic eye muscles. There are six of these fulfilling identical functions in all gnathostomes. Four rectus muscles rotate the eyeball upwards, downwards, forwards or backwards. Two oblique muscles, situated more anteriorly, turn it respectively upwards or downwards and forwards. The anterior, superior and inferior recti, and the inferior oblique are derived from the first myotome on each side—the myotomes of the premandibular segment—and are innervated by the third cranial nerve, the oculomotor. The superior oblique is derived from the myotome of the second, or mandibular, segment, and the posterior rectus from that of the third segment. They are innervated respectively by the fourth and sixth nerves—the trochlear and abducens. Myotomes are missing from the otic region and those that are situated dorsal to the gills cannot grow vertically downwards, as do more posterior ones, because of the presence of the gills. Therefore, although the epaxial musculature in front of the pectoral girdle, the epibranchial musculature, retains its myomeric arrangement, most of the hypaxial musculature in this region has been replaced by branchial musculature. The persisting hypaxial musculature comprises processes that pass posteriorly, ventrally and then anteriorly to give the hypobranchial musculature. It has lost its myomeric arrangement, is restricted to the ventral region of the body, and is overlain by branchial muscles. The coracomandibular opens the mouth and the rectus cervicis expands the oropharyngeal cavity.

(c) Trunk myotomes

i The overall organization of the trunk myotomes was referred to above. In the dogfish each one contains slow, fast and superficial fibres with the first two types being further subdivisible giving a total of 5 fibre types. The general arrangement is shown in Fig. 3.13. The outer border of the myotome consists of a single, sometimes interrupted, layer of superficial fibres that are not present in all sharks and are absent from rays. Internal to these are the so-called Type I slow fibres which merge internally into a second type of slow fibre of larger diameter. There is then a boundary with Type I fast fibres and these merge into larger, Type II, fast fibres.

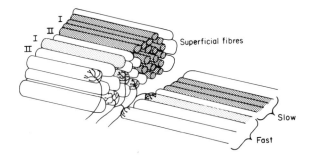

Fig. 3.13 Dogfish myotome seen from the internal aspect. Superficial fibres overlie slow fibres. Fast fibres are innervated terminally. (After Bone.)

The fast fibres differ from the slow ones in the virtual absence of mitochondria, a thinner Z line and a more abundant sarcoplasmic reticulum. Each fast fibre is innervated by two separate axons and in some elasmobranchs the motor endplates associated with these can be distinguished by vesicle content. Direct electromyographic studies on spinal shark preparations demonstrate that electrical activity occurs in the superficial fibres during slow swimming and in the deeper fast fibres during rapid swimming. Furthermore, the slow fibres largely operate by aerobic glycolysis and lipolysis, fast fibres by anaerobic glycolysis.

ii The development of the median fins involves, first of all, a narrow ridge of epidermis. This is the precursor of a wider longitudinal fold into which mesoblastic mesenchyme penetrates, giving a provisional embryonic fin of some length. Local thickenings soon appear within this and it is these regions alone that develop into the adult fin. The intervening parts of the fold disappear. In each fin rudiment a thick longitudinal plate of mesenchyme is the precursor of the future fin skeleton. On either side of this plate myotomal derivatives grow into the fold to give the radial muscles (Fig. 3.14). Each myotome contributes one muscle bud from its dorsal edge but as the adult fins are considerably concentrated at their base, the body length increasing faster than the base of the fin, the more anterior and posterior myotomal derivatives appear to move away from their source to reach the base of the fin. They also decrease in size at the two ends but each adult radial muscle corresponds to one segment.

iii The development of paired fins follows a similar course. They are first indicated by a narrow ridge of folded epidermis which subsequently grows out as a longitudinal fold of the body wall into which grow muscle buds from the neighbouring myotomes. They, too, exhibit the basal concentration noted in connection with median fins. This origin is reflected in the adult by the innervation from successive spinal nerves in a regular sequence. It was the embryological origin from an extensive fold which suggested the '*finfold theory*' for the origin of paired fins during phylogeny. They were considered to be the remnants of continuous lateral fin folds, themselves forward extensions of the medial fold from which caudal and anal fins are derived. This was envisaged as having divided just behind the anus and run forwards as continuous lateral folds.

In the adult the gross muscle associations of the pectoral fins are single dorsal and ventral masses. That on the dorsal side originates from the pectoral girdle, is inserted on the dorsal surface of the pterygiophores, and functions as a levator of the fin. An antagonist, the pectoral depressor, is inserted on the ventral surface of the pterygiophores. Contraction of particular parts of these muscles also moves the fin forwards—protraction—or backwards—retraction.

(d) The branchial musculature

i The branchial musculature develops from embryological aggregations that give rise to several muscles. It is assumed that, primitively, each visceral arch of some hypothetical ancestor possessed an identical set of these and the fact that this is not the case in any living gnathostome is the result of differential specialization and loss.

ii The cucullaris (Fig. 3.15) originates from the surface of the epaxial musculature and is inserted on both the pectoral girdle and the seventh, last, visceral bar. The five large constrictor muscles lie ventral to it, are numbered 2–6 according to the visceral arch with which they are associated, and each is rather arbitrarily divided into dorsal and ventral components. Acting together, these dorsal and ventral components compress the gill chambers and aid the expulsion of water from the gill slit.

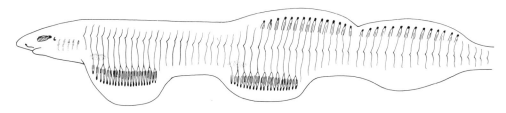

Fig. 3.14 Diagram showing the origin of the radial muscles of fins from myotomal outgrowths. They are innervated by spinal nerves of the appropriate segments.

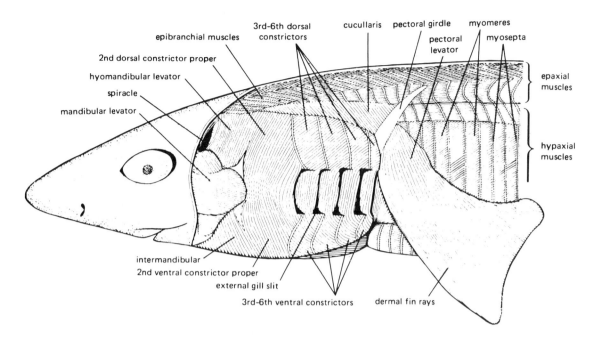

Fig. 3.15 Musculature of *Squalus* seen from the side. (From Holmes.)

iii The constrictor of the mandibular arch is modified. A large mandibular levator originates from the palato-pterygo-quadrate, is inserted on Meckel's cartilage, and closes the jaw. Broad transverse intermandibulars which are situated on the ventral side of the head behind the lower jaw originate on Meckel's cartilage and are inserted on a longitudinal mid-ventral raphe. Their simultaneous contraction elevates the floor of the buccal cavity. The second dorsal constrictor is broader than the subsequent ones, originates on the skull and epaxial musculature and is inserted on the hyomandibular forming the hyomandibular levator. Its ventral analogue is also broad and its anterior components form the interhyoideus which helps to elevate the floor of the orobranchial cavity. The third, fourth and fifth dorsal constrictors are similar to each other, originate partly on the cucullaris and partly on a vertical raphe, and are inserted on the ventral constrictors and a branchial bar. Their ventral homologues are also all similar, originate on the hypobranchial musculature and a vertical raphe, and are inserted on the dorsal constrictors and a branchial bar. The dorsal and ventral components of the sixth constrictor originate, respectively, on the cucullaris and pectoral girdle, and the pectoral girdle and fin musculature. A coracobranchial lies deep to the constrictors, originates on the pectoral girdle and

inserts, by separate components, on the visceral bars 3–7. An anterior element is closely associated with the rectus cervicis. The action of the muscle helps to expand the orobranchial chamber. Although innervated by the hypobranchial nerve it develops with the branchial musculature during ontogeny.

ELECTRIC ORGANS

i An electric organ consists of electroplaxes. These lie in columns which run dorsoventrally in the electric rays and are derived embryologically from electroblasts that are similar to the sarcoblast precursors of muscle. The elongate multinucleate cell enlarges laterally, nuclei become distributed at the cell periphery, and the centre becomes filled with mucous protein. One surface, usually the electronegative one, is innervated, and the other is papilliform. The whole structure, which resembles a motor end plate, is a modified neuromuscular junction.

ii In *Torpedo marmorata* each electric organ has about 450 columns each with 400 electroplaxes, and the electric tissue represents one-sixth of the body weight. The plates in each column are in series, the various columns are in parallel, and the whole is innervated by huge branches of the vagus (X). During discharge, each organ fires repetitively for 3–5 spikes, each spike being of uniform height and 3.5 ms duration. Both organs discharge within 0.1 ms of each other and this synchrony is essential for maximum power output. The total peak voltage measured in the absence of external resistance is 20–30 volts (cf. *Electrophorus* page 54).

LOCOMOTION

i Fish vary greatly in their appearance and their methods of locomotion are correspondingly diverse. Species with a sharklike body have been most extensively analysed but even these are heterogeneous and the factors operating in other cases can be quite different. In a 'typical' pelagic fish the development of tension in the muscles on one side of the body tends to rotate both the head and tail towards that side. Owing to its large size the anterior end only moves laterally to a limited extent but the tail is swept towards the side contracting. The forces operating on the tail have both lateral and forward components. As it moves to the left the resultant force acts forward and to the right, and vice versa. Thus, during a complete cycle of tail movements the forces to left and right cancel out, whilst the forward components overcome the drag exerted on the body by the external medium and propel the fish forwards.

ii All shark-like Chondrichthyes swim by means of their lateral muscles but the actual method varies from such a simple side-to-side movement of the tail, to the passage of waves posteriorly along the entire body driving water backwards and the fish forwards. During forward locomotion the fish also needs to maintain its position in the water, or control its rate of change in depth, and to counteract any tendency to pitch, roll or yaw. Sharks and dogfishes, whose specific gravity is greater than that of water, have a characteristic heterocercal tail. The posterior end of the vertebral column turns upwards and there is a large lower flap on the asymmetrical caudal fin. As the tail sweeps from side to side its leading surface is directed obliquely downwards and backwards so that the water exerts an upward pressure that tends to raise the tail during motion. This upward force at the tail tends to depress the head end and the tilting effect about the centre of gravity is counteracted by the set of the pectoral fins against which the water also exerts an upward pressure (Fig. 3.16). These are held with their hinder sides obliquely downwards. The combined effect of the forces on the pectorals and tail is therefore to raise the fish vertically

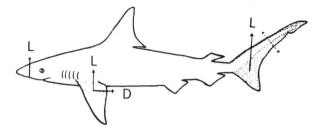

Fig. 3.16 Diagram showing the forces acting on a shark during locomotion.

in the water, and during locomotion at constant depth these two sets of upward forces exactly balance the body weight. In broad terms, if the fish stops swimming it will begin to sink. Obviously, such a facility to maintain a uniform depth would be of the utmost importance to species such as the whale shark, basking shark, mackerel shark, thresher shark, blue shark or white-tip shark which range over the deep ocean. It is of less significance to dogfish or ground sharks that inhabit shallow waters and can therefore rest on the sea-floor. In fact, not all selachians are substantially denser than sea water. In some, such as *Cetorhinus* the basking shark mentioned above, the presence of lipids and the hydrocarbon squalene, particularly in the liver, means that their density is very close to that of sea water. Their pectoral fins do not have to produce such appreciable vertical forces in order to maintain constant depth and their importance is principally restricted to manoeuvring. The pelvic fins in all these fish are held parallel to the longitudinal axis of the body and play little part in such controls. The differential development of the muscles of the two sets of paired fins, and of the pectoral and pelvic girdles, is, therefore, intelligible.

iii The set of the pectoral fins also controls pitching movements and by an alteration of their angle the fish can turn upwards or downwards. Any rolling movement, or rotation around the longitudinal axis of the body that occurs during forward locomotion can be controlled by setting the pectoral fins on the two sides at different angles. The side of the body on which the upward force is greatest will then be forced upwards. In the case of yawing, turning to right or left of the direction of movement, the use of experimental models shows that if the fins in front of the centre of gravity are removed the fish remains stable to yawing. However, if all the fins except the caudal are removed it is unstable and any small deviation results in it turning further from the original trajectory. The caudal, posterior dorsal and ventral fins prevent the animal from oscillating to the right or left of its path like the vertical flight of an arrow or the vertical rudder of a plane.

iv In many fishes, flexures of the body or tail are not the principal means of locomotion. At one extreme the tail and its fin are virtually useless in this respect. This is evident in many rays (Batoidea). In two major groups, the saw-fishes and guitar fishes, the trunk and tail, which are similar to those of a shark in form, are flexed and sculled in a shark-like manner. Undulations of the pectoral fins are, at most, an auxiliary method of propulsion. Sculling motions of the tail, carried out in a somewhat lazy way, also propel the electric rays, but in the skates, sting-rays, eagle-rays, cow-nosed-rays and devil-rays the tail is reduced to a trailing appendage. Not only are the pectoral fins very large and packed with serial muscles moving corresponding radial cartilages, but there is a close correlation with their way of life.

Skates and sting-rays spend much of their time on the sea-floor. When they move, undulations follow one another down the extensive pectoral fins with the wave height increasing from the front to the middle of the disc and decreasing thereafter. These travelling waves push against the water and generate the required thrust. The eagle-rays, cow-nosed-rays and devil-rays lead a more active, free-swimming existence and all are capable of moving swiftly through the sea by graceful flapping movements of the pectoral fins. The Holocephali also contrast with shark-like genera. They swim by sculling the body along with their pectoral fins.

DIGESTIVE SYSTEM

i The alimentary canal is a relatively uncomplicated tube with an S-shaped curvature about midway along its length. The oral cavity contains the immovable tongue supported by the hyoid arch. Whether the teeth are adapted for biting or crushing, they are all uniform in appearance, are homodont, and have no roots, but they are closely united to the connective tissue covering the jaw cartilages, are acrodont. They are also arranged in rows with each functional tooth having a reserve series behind it so that if it is destroyed a reserve tooth replaces it. This can recur an indefinite number of times, a condition known as polyphyodont.

ii The more posterior pharynx bears the internal gill openings. The first pair, the spiracles, are rounded and open to the dorsal surface in skates—the lateral surface in sharks. A pseudobranch on their anterior border consists of folds of mucous membrane homologous with the gill filaments of more posterior gills. There are usually five pairs of these last, opening laterally in sharks and dogfish but ventrally in skates and rays, but six pairs occur in *Hexanchus* and *Pliotrema*, and *Heptanchus* (*Heptranchias*) has seven pairs.

iii The sometimes constricted posterior end of the pharynx leads into the oesophagus which, unlike the situation in lampreys, enlarges to form the U- or J-shaped stomach (Fig. 3.17). The initial, cardiac, segment

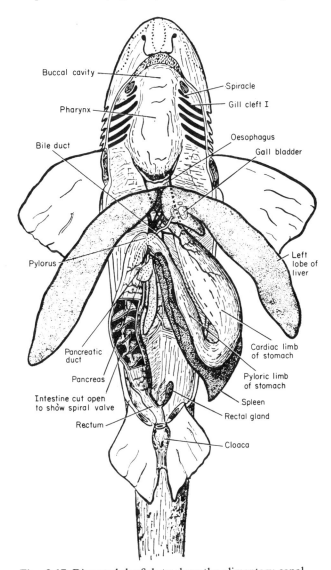

Fig. 3.17 Dissected dogfish to show the alimentary canal.

is large and bears backwards-directed papillae internally. The shorter pyloric segment terminates in a constriction and muscular thickening—the pylorus or pyloric sphincter. This marks the beginning of the duodenum. The secretions of the pharynx are essentially all mucous but the stomach has acid-secreting cells whose secretions prevent bacterial decay.

iv The duodenum is a short tube connected posteriorly with the remainder of the small intestine. This is a rather wider, straight structure whose external surface bears a spiral line marking the attachment of a spiral valve in the lumen. Absorption takes place throughout the entire intestine. A finger-shaped rectal gland opens into the intestine and is responsible for the excretion of large amounts of sodium chloride from the blood. It is degenerate in the Lake Nicaragua shark, a sub-species of *Carcharinus leucas*, because salts tend to diffuse out of the animal not into it as is the case in marine forms.

v The liver is a trilobed structure and the greenish gall-bladder is associated with the median lobe. The bile duct, or ductus choledochus, leaves the gall bladder and passes through the mesentery that supports the liver and gut—the gastro-hepatic omentum. The whitish pancreas has a larger dorsal and smaller ventral lobe that are joined by a bridge of tissue near the pylorus.

RESPIRATORY SYSTEM

i Water taken into the mouth passes through the last three pairs of internal gill slits and their parabranchial chambers, bathes the gill lamellae, and passes out through the external gill slits. Water taken in by the spiracle passes out through the first three gill slits of that side. Gaseous exchange occurs at the gills. Except for the spiracular pseudobranch which receives oxygenated blood, these are supplied with deoxygenated blood which takes in oxygen from, and gives off carbon dioxide to, the water bathing them. They comprise rows of lamellae which radiate from the membranous covering of the cartilaginous visceral arches and are supported by numerous cartilaginous gill rays extending from the arches into the interbranchial septa. Gill rakers project into the pharynx from the inner surfaces of all these visceral arches. The gill lamellae themselves are unlike those of cyclostomes as they are closely attached to the interbranchial septa which extend laterally past the outer edges of the lamellae and are continuous with the skin, thereby contributing to flaps that protect the external gill openings.

The hyoid arch in front of the first typical gill slit bears gill lamellae on its posterior surface only—a hemibranch. An entire gill or holobranch is made up of two hemibranchs with an interbranchial septum between them. In *Squalus acanthias*, for example, there are nine hemibranchs but only four holobranchs on each side, the last gill pouch bearing lamellae only on its anterior wall. Nevertheless, there are five cartilaginous visceral arches behind the hyoid as no lamellae are associated with the last one.

ii Some sharks do not make breathing movements but simply swim with their mouths open. Where breathing movements do occur, the anterior part of the body first enlarges and then gets shallower and narrower as the mouth cavity and parabranchial chambers contract. Both the mouth and gill slits are open during the intervals between breaths when the mouth cavity and parabranchial chambers are expanded. The mouth and spiracle then close while the gill slits remain open as the cavities contract and water is expelled from the system. There is, at this moment, a positive pressure in both the mouth and parabranchial chambers. The mouth and spiracles then open again, and the gill slits close briefly, as the cavities expand and water is drawn in. A constant flow of water from the mouth to the parabranchial chambers is maintained throughout the cycle as the pressure is always higher in the mouth. The movements themselves reflect the integrative activity of respiratory neurons in the medulla oblongata and differential contractions of the relevant musculature. During normal, gentle breathing, muscles such as the coracomandibular, coracohyoideus and coracobranchial are inactive and there are three main phases. The levator mandibulae contracts and closes the mouth. It then remains contracted as the constrictors contract from the front to the back and force water out through the external gill slits. All muscles then relax and water is drawn in.

URINOGENITAL SYSTEM

(a) Kidney

i The majority of selachians are marine and their total osmotic concentration is somewhat above that of sea water. This is partly due to the presence of inorganic salts but particularly to the presence of urea and trimethylamine oxide both of which are present in unusually high concentrations. As a result, there is some tendency for water to diffuse into the body, particularly at the gills. Inorganic ions also tend to diffuse in while urea and trimethylamine oxide tend to diffuse out. The gill lamellae are largely impervious to urea and trimethylamine oxide, which are therefore retained, but ammonia is lost. In animals like *Pristis*, which penetrates up the Mississippi and Chinese rivers, the concentration of urea in the blood increases the amount of osmotic work that is necessary. The principal excretory organs controlling the overall levels are the kidneys but there are cells in the gills which may excrete salts, and the rectal gland excretes sodium chloride.

ii The kidney of an adult selachian is a mesonephros homologous with, and analogous to, that of cyclostomes. Here, however, the excretory and genital systems are closely associated. In all gnathostomes, instead of there being a single duct on each side there are two. One of these, the Wolffian duct, can be regarded as the original mesonephric (pronephric) duct and continues to receive the tubules from the Bowman's capsules. The other is the Mullerian duct. This opens from the coelom by the remnants of the degenerated pronephric tubules, leads straight back to the cloaca and lacks any connection with the mesonephric tubules. The urinary functions of the mesonephros are limited to its hind regions, sometimes referred to as an opisthonephros. This consists of a mass of glomeruli and tubules. A section of each tubule is involved in urea absorption and they join to give a series of five urinary ducts which join the Wolffian duct and enter a urinogenital sinus which, although of mesodermal not endodermal origins, functions like the tetrapod bladder.

(b) Male genital system

i In the male the testes are paired and connected by their anterior ends to the mesonephric tubules via vasa efferentia. These correspond to the original coelomostomes, and through them the sperm gain access to the Wolffian duct which becomes known as the vas deferens and has a swollen hind end—the seminal vesicle. The anterior regions of the mesonephros are, therefore, concerned with the evacuation of male genital products (Fig. 3.18).

ii Fertilization is internal. The posterior ends of the male's pelvic fins are modified as 'claspers' which function as a penis. Each is cigar-shaped, stiffened by cartilages known as rhipidion, claw, spur and distal cartilages, and bears a deep groove along one side. They normally lie flat against the body with their apices

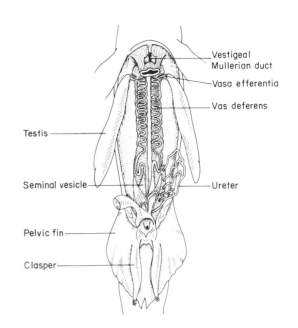

Fig. 3.18 Reproductive system of a male dogfish.

directed backwards but can be erected by contraction of dorsal, medial and ventral flexor muscles. At the same time the contraction of a dilator muscle spreads the distal cartilages. The male approaches and, in some cases, wraps his body around the female and inserts one clasper into her cloaca. The spreading of the distal cartilages presumably anchors it in her. When the clasper is erect the end of the groove is against her cloaca and squirts semen into it. This is assisted by the expulsion of water from siphon sacs that lie on the ventral side of the animal and open into the clasper grooves. Each can be charged with water by repeated erection of the ipsilateral clasper.

(c) The female genital system

In females the mesonephros is wholly excretory in function. The right ovary is large and suspended in the coelomic cavity by a fold of coelomic epithelium—the mesovarium. The left ovary disappears. When ripe, the large eggs drop free into the coelom, enter the funnel-shaped openings of the Mullerian ducts (Fig. 3.19) on the ventral side of the oesophagus, and pass down them. These oviducts swell out forming oviducal glands, which secrete the horny egg case, and then open separately into the cloaca. Internal fertilization permits the enclosure of the eggs within this egg case, as in

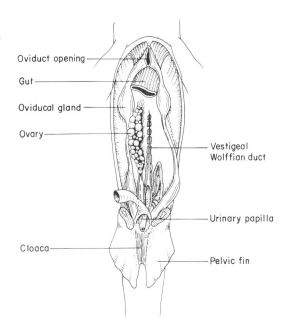

Fig. 3.19 Reproductive system of a female dogfish.

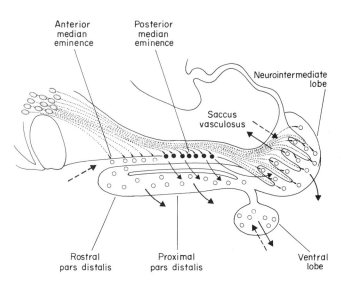

Fig. 3.20 The selachian pituitary; a diagrammatic sagittal section showning hypothalamo-hypophysial vascular and neurosecretory links. Neurosecretory cells (top left) in hypothalamus (preoptic nucleus) send axons to median eminence and neurointermediate lobe. Solid circles, primary portal capillaries; open circles, secondary portal capillaries and other intra-hypophysial capillaries; broken arrows, arteries; solid arrows, veins. (After Holmes & Ball.)

dogfishes, but it also makes it possible for the eggs to be retained within the mother. This occurs, for example, in the spiny dogfish, *Squalus acanthias*, the nursehound, *Mustelus canis*, and the sting-ray, *Dasyatis*. The first-named retains the developing young for nearly two years. At first they are bathed by a fluid resembling blood plasma but later by one more resembling sea water. The nursehound only retains its young for 10–11 months and, as the yolk is used up, part of the yolk sac adheres to the uterine wall forming a placenta-like structure for transfer of nutrients and for the excretion of waste products. *Dasyatis* lacks such a placenta but the uterus contains a maternal secretion containing 13% organic matter, largely in the form of fats.

ENDOCRINE ORGANS

(a) The hypophysis

i The hypophysis is a conspicuous structure attached, as always, to the ventral diencephalon (q.v.). The adenohypophysis comprises a very large pars intermedia, that is intermingled with the neurohypophysis to form a neurointermediate lobe, and an elongated pars distalis, itself comprising dorsal and ventral lobes (Fig. 3.20). In sharks and some rays the entire pars distalis is hollow with the anterior part of the dorsal lobe

containing vesicles or tubules communicating with the central hypophysial cavity, and the posterior part consisting of folds of tissue around the cavity. In other skates and rays the cavity is reduced or obliterated and the dorsal lobe is a compact mass of cords or clusters of cells. The pars intermedia is located below a thin neurohypophysial layer and penetrated to a variable extent by neurohypophysial fibres giving the neurointermediate lobe. The constituent intermedia cells can be organized as distinct lobules separated by highly vascular connective tissue, or as a mass of cells with an irregular blood-vascular plexus. They secrete melanophore-stimulating hormone. The secretion of adrenocorticotrophic, gonadotrophic and thyrotrophic hormones has long been known and it is clear that, taken together, the dorsal and ventral lobes fulfil the functions of the pars distalis.

ii Neuro-secretory fibres of the preoptico-hypophysial tract enter the neurointermediate lobe. In *Squalus acanthias* they terminate in a fairly well-defined pars nervosa but in other species they are more diffuse. In *Squalus* most terminate on an outer layer of ependymal fibres which follows the boundary of the intermedia.

Elasmobranchs, generally, possess arginine vasotocin and one or two oxytocin-like factors but little is known of their functions.

(b) The thyroid gland

The thyroid is a rather dark flattened mass in the floor of the pharynx often just posterior to the mandibular symphysis. It frequently reflects its phylogenetic origins —remaining attached to the pharyngeal floor by a narrow stalk with a residual ciliated pit recalling the endostyle. After injections of the radioisotope I^{131} maximal uptake is detected at 17 hours and the concentration continues to increase for four days. There is no clear respiratory response following thyroidectomy but it is clear that the gland is under the control of thyroid-stimulating hormone (TSH) secreted from the ventral lobe of the hypophysial pars distalis.

(c) The ultimobranchial glands

The ultimobranchial glands are calcium-regulating glands which develop during embryonic life from the epithelium of the last gill pouch and, as such, have a similar origin to parathyroids (Fig. 3.21). In *Squalus acanthias* the gland occurs, on the left side only, between the pericardium and the ventral surface of the pharynx. It is highly vascularized and consists of many follicles containing two types of epithelial cells. Both types

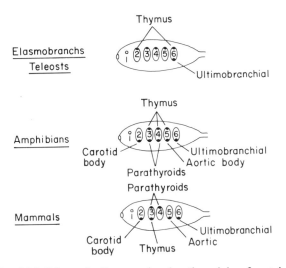

Fig. 3.21 Schematic diagram showing the origin of certain branchial derivatives in different classes. (After Copp.)

contain small membrane-enclosed granules in their apical and basal regions. The secretory product is calcitonin and its concentration is far higher than that in mammals. All vertebrates have some capacity to regulate their serum calcium at a level different from that in the environment and in *Carcharinus leucas* the ionic calcium is 3.1 mmol l^{-1}. In the freshwater subspecies from Lake Nicaragua the value is 1.7.

(d) The inter-renal bodies

The inter-renal bodies situated between the posterior lobes of the kidneys are the selachian homologues of the mammalian adrenal cortex. Experimental manipulations demonstrate that they control the rate of secretion of sodium chloride from the rectal gland (q.v.) and, as in other vertebrates, have the general effect of maintaining the ionic concentration of the body. It has been suggested that they are coelomostome derivatives which retain an ancient function in modified form.

(e) Chromaffin bodies

These are the selachian homologues of the mammalian adrenal medulla. They are small, paired masses lying on the intersegmental branches of the dorsal aorta. Some of the more anterior members may be elongated and united. There is frequently a clear association with the sympathetic ganglia (q.v.).

(f) The urophysis

The urophysis is a caudal neuro-endocrine apparatus that is a characteristic feature of the posterior spinal cord in both elasmobranchs and teleosts, and is very extensive in the first-named. Its secretory cells, modified neurons, project to a rich capillary network where their terminal axonal enlargements contain secretory products. The vasopressor and antidiuretic properties of extracts suggest analogies with neurohypophysial factors (cf. Fig. 3.22).

(g) The endocrine pancreas

The compact pancreas of selachians contains both exocrine and endocrine tissue. The latter occurs mainly as an outer layer along the epithelium of small ducts, or as single cells at the base of large ducts. The presence of A, B, D, and granular cells suggests the presence of insulin, glucagon, and at least one other hormone. Insulin has long been known.

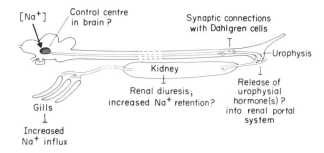

Fig. 3.22 Diagram showing possible role of the urophysis in osmoregulation. (After Fridberg & Bern.)

THE CARDIOVASCULAR SYSTEM

i There is a similar basic organization underlying the cardiovascular system of all vertebrates and that of selachians, therefore, has much in common with that of cyclostomes. The heart, in its pericardium, is bent on itself to form an S shape. The sinus venosus, which receives the ductus Cuvieri and the hepatic sinus from the liver, opens into a large, dorsally placed atrium which, in turn, opens into the ventricle. When the atrium contracts there is a slight positive pressure within it whilst that in the ventricle is about zero. The ventricle then drives blood into the muscular conus arteriosus which has two rows of valves preventing the blood from returning. The blood is driven into the conus and aorta at pressures of about $30-40$ cm H_2O ($3-4$ kN m^{-2}) and the conus continues to contract as the ventricle relaxes so that a high pressure is maintained—that in the elastic ventral aorta only fluctuating between $30-40$ cm H_2O. The elastic properties of the ventral aorta are due to the high proportion of elastin (31%) in its walls compared with only 9% in the dorsal aorta.

ii Afferent branchial arteries pass from the ventral aorta to the gills and breakdown into the capillaries of the gill lamellae in the hyoid and four branchial arches. The oxygenated blood is collected into four efferent branchial arteries corresponding to gill slits 1–4. These lead to the dorsal aorta and in selachians, as in Dipnoi, each is made of two collecting vessels, one on each side of the gill slit and uniting above it. The gill arches, therefore, contain two efferent vessels and one afferent one (Fig. 3.23). The hemibranch of the 5th gill slit is drained by the fourth efferent. The spiracular pseudobranch receives oxygenated blood from the anterior hemibranch of the 1st gill slit and an efferent pseudo-

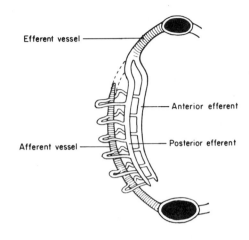

Fig. 3.23 Diagram illustrating the derivation of the adult branchial vessels of selachians from the continuous embryonic arch (hatched). (After Goodrich.)

branchial artery leads from it to the internal carotid artery which is an extension of the dorsal aorta and supplies the brain inside the skull.

iii The dorsal aorta develops from paired rudiments which remain separate in the branchial region and head but fuse below the notochord to form a median dorsal aorta behind. This last gives off paired subclavian arteries to the pectoral region and passes back to send coeliac, lieno-gastric, anterior and posterior mesenteric arteries into the dorsal mesentery and thence to the stomach, liver, intestine and well-developed spleen. Further back it gives off pelvic arteries to the pelvic girdle region, renal arteries to the kidneys, as well as branches to the gonads, and continues into the tail as the caudal artery.

iv During passage through the capillary bed of the gill lamellae the blood pressure is greatly lowered and it is probably for this reason that the venous return in selachians involves a system of large sinuses—the resistance offered by a blood vessel decreasing with an increase in diameter. Blood from the tail returns forwards via the caudal vein which forks to send a renal portal vessel to each of the kidneys. Here the blood passes through a network of capillaries around the tubules (Fig. 3.24). It is then transported forwards by the posterior cardinals to a posterior cardinal sinus and thence via the ductus Cuvieri to the heart. The blood from the digestive tract and spleen is drained by the hepatic portal system which converges on the liver and

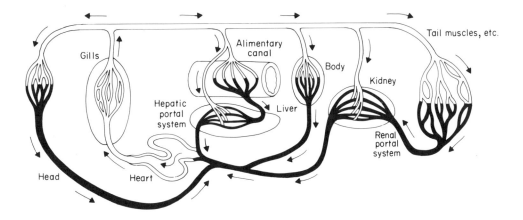

Fig. 3.24 Diagram of the vascular system in a fish—
arteries white, veins black.

then the hepatic vein runs forward to the sinus venosus. Blood from anterior regions returns via the inferior jugular and anterior cardinal sinuses. The inferior jugular drains the floor of the mouth, the region ventral to the visceral arches and the pericardial region. The anterior cardinal sinus lies dorsal to the visceral arches and receives blood from the post-orbital sinus between the hind border of the orbit and the first hemibranch. A hyoidean sinus in front of the hyoidean hemibranch connects the inferior jugular and anterior cardinal sinuses.

v The spleen, which is a haematopoietic organ, forms a prominent component of the abdominal viscera. It is a dark triangular structure loosely attached to the posterior bend of the stomach, at the point where the cardiac and pyloric portions join, by means of the gastrosplenic omentum.

THE SENSE ORGANS

(a) Olfaction

Two nostrils open on to the ventral side of the snout and lead into a blind pouch—the olfactory sac. An oro-nasal groove may connect each one to the mouth and a flap of skin over the nostrils directs water in one side and out of the other. Connections with the mouth also frequently permit the respiratory current to draw water through the olfactory region. Within the olfactory sac there are many small lamellae covered by olfactory epithelium in which lie numerous fibres of the olfactory nerve. These synapse with fibres in the olfactory bulb. Nurse sharks,

Ginglymostoma, certainly use gradients of intensity to locate the source of odours and this is presumably a widespread phenomenon in the location of prey.

(b) Photoreceptors

i Ultrastructural investigations confirm that the saccular pineal organ is again a specialized photoreceptor whose component cells bear a strong morphological resemblance to retinal receptors. The paired eyes are also very well developed and retinoscopic tests show that the animals are hypermetropic or long-sighted. Moreover, they are more hypermetropic in the lateral parts of their visual field than in the forward components which, instead, have the greatest visual acuity. Hammer-head sharks have no binocular field of vision.

ii The eyeball is made up of outer sclerotic, and internal choroid layers. A supra-choroid vascular layer intervenes on the medial aspect in those species which possess a cartilaginous pedicel amongst the rectus muscles. This pedicel helps to hold the eyeball in place. Towards the external surface of the eye the choroid coat and non-sensory region of the internally situated retina separate from the sclerotic and form a muscular, curtain-like structure—the iris (Fig. 3.25). Closure of the iris cannot be achieved by the use of autonomic drugs. It is an independent effector and closes when illuminated, even following isolation of the head. The radial dilator fibres, which open the pupil, are, however, innervated by the oculomotor nerve, and do respond to drugs. In diurnal predators the iris divides the pupil into two slits by means of an upper flap or operculum.

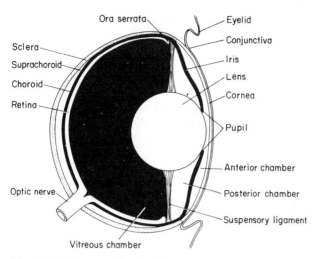

Fig. 3.25 Diagram of a dogfish eye

Labels on figure: Ora serrata, Eyelid, Sclera, Conjunctiva, Suprachoroid, Iris, Choroid, Lens, Retina, Cornea, Pupil, Optic nerve, Anterior chamber, Posterior chamber, Suspensory ligament, Vitreous chamber

Behind it is a series of radiating folds, the ciliary body, to which the crystalline lens is attached by a suspensory ligament. The lens is spherical, very hard, and provided with a protractor lentis muscle which swings it forward for accommodation during near vision. The space between the iris and cornea is the anterior chamber. The smaller space between the iris and lens is the posterior chamber. Both are filled with aqueous humour. The large cavity behind the lens, the vitreous chamber, is filled with a jelly-like mass—the vitreous humour. The sensory portion of the retina, bearing rod cells, extends as far as the ciliary body and the irregular border of this region is the ora serrata. The point of exit of the optic nerve results in a blind spot.

(c) The lateral line system

i The neuromasts of cyclostomes all protrude from the surface of the body but, in contrast, those of elasmobranchs are enclosed within sunken canals. Along these canals the neuromasts alternate with openings to the external medium and a difference between the external pressure at two successive pores displaces the water in the canal, thereby deflecting the neuromast cupula. Superficial neuromasts are extremely sensitive to water movements along the surface of the body but this is not the case with those that are sunk in canals as the pressure differences between two successive pores are inadequate. Masking effects arising from the water streaming over the body during locomotion are, therefore, reduced and this enhances the sensitivity to gentler water movements in other directions. These may be generated by the movements of potential prey or by the approach of predators.

ii The lateral line system of sharks and dogfishes consists of two lateral canals that run the length of the body from head to tail and overlie the lateral septum separating epaxial and hypaxial myotomal derivatives. At about the level of the spiracle they are connected by a transverse supratemporal canal and anterior to this they both branch. A supraorbital canal runs dorsal to the eye; an infraorbital one, having given off a hyomandibular branch which passes to the hyoid arch, runs forward below the eye and then to the ventral surface of the snout. A mandibular canal lies in the lower jaw. The neuromasts of the definitive lateral line canals are innervated by the lateral line branch of the vagus nerve, those of the cephalic region by branches of the facial and glossopharyngeal nerves.

(d) The ampullae of Lorenzini

Scattered over the cephalic region of selachians there are also numerous small pores which represent the openings from ampullae of Lorenzini. Each leads to a small jelly-filled canal which terminates in the bulb-like ampulla itself. Sensory cells situated within this are innervated by about six axons of the facial nerve, and transection of these abolishes responses to electric fields that resemble those around moving prey. Under constant conditions action potentials in these axons occur at a steady rate. Changes of temperature, salinity and external electric potentials affect this rate and it is particularly sensitive to potential gradients along the length of the ampullae. The low electrical resistance of the jelly, coupled with the high resistance of the ampullary walls, leads to electric currents being channelled along the ampulla.

(e) The ear

i Fish lack external and middle ears and have only an inner ear which is primarily concerned with detecting the animal's orientation relative to gravity and maintaining balance. The overall size of the organ varies as a function of body size but it always comprises a more or less delicate membranous labyrinth embedded in a cartilaginous labyrinth. Anterior, posterior and horizontal semicircular canals each have an ampulla at their lower end together with a vestibule made up of saccular and utricular regions. In *Squalus* the anterior and horizontal canals join a delicate chamber, the anterior utriculus, dorsally. Their closely apposed ampullae also

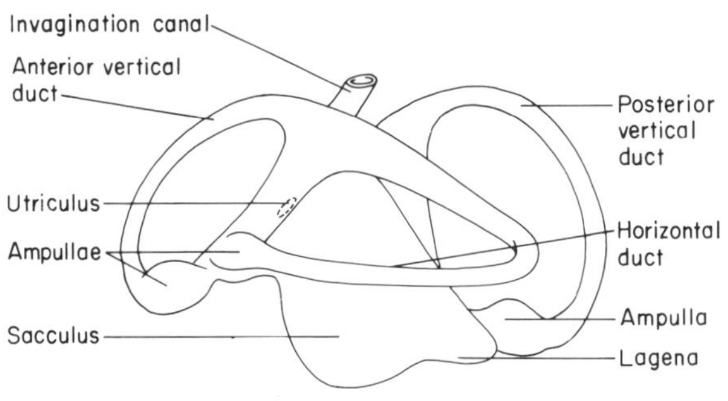
Invagination canal
Anterior vertical duct
Utriculus
Ampullae
Sacculus
Posterior vertical duct
Horizontal duct
Ampulla
Lagena

Fig. 3.26 Lateral view of left ear of a selachian.

join it ventrally. Both ends of the posterior canal join another chamber considered to be a second or posterior utriculus. The sacculus is a pear-shaped structure between the two utriculi. It opens to the top of the head by an 'invagination' canal and a blunt projection, the lagena, extends posteriorly from its ventral side (Fig. 3.26). The cavities of the sacculus and utriculi intercommunicate by small apertures.

ii The membranous labyrinth is filled with endolymphatic fluid which, in elasmobranchs, is largely sea water. Sand grains may enter the sacculus from the exterior, via the invagination canal, and these, together with endogenous crystals of calcium salts, are combined into the loose mass of the otolith. Movements of both the endolymph and otolith lead them to impinge upon sensory receptors and thereby give information about the direction of gravity. These receptors are akin to the neuromasts of the lateral line and their spontaneous discharge rates are increased or decreased by displacement in opposite directions.

(f) Taste buds

Taste buds are widely distributed in the oro-pharyngeal region and reported to detect the same qualities—salt, sweet, sour and bitter—as ours do.

THE NERVOUS SYSTEM

(a) The cranial nerves

i It will be clear from the discussion of the ammocoete larvae of lampreys (page 6) that the interpretation of the cranial nerves is of prime importance for an understanding of the segmental origins of the cranial region. The classical anatomists gave these nerves specific names and numbers and it was only later that the underlying metameric pattern was appreciated. The anterior nervus terminalis, where it occurs, is presumably a prostomial component derived from a region in front of the metamerically segmented part. The olfactory nerves also differ from the rest, as their component fibres arise from cells in the nasal mucosa and grow back to the forebrain, and the optic nerve is really a part of the brain.

ii The segmental derivation of the remaining nerves is summarized in Table 3.1 and represented in Fig. 3.27. The dorsal root of the first or premandibular segment is the profundus. In modern forms this appears to be a branch (ramus) of the trigeminal nerve and in *Squalus*, *Mustelus*, skates and rays traverses the orbit, gives off the long ciliary nerve to the eyeball, passes between the oblique muscles, and finally leaves the orbit for the superficial regions of the snout. The corresponding ventral root is the oculomotor which serves the premandibular myotomal derivatives—the superior, inferior and anterior rectus muscles and the inferior oblique muscle.

Table 3.1 The underlying metameric segmentation of the cranial nerves and cranial region.

Segment	Arch	Dorsal root	Ventral root
1. Premandibular	Trabecula	Profundus (often cited as a ramus of V)	Oculomotorius III
2. Mandibular	Palato-pterygo-quadrate bar and Meckel's cartilage	Trigeminal V	Trochlear IV
3. Hyoid	Hyoid	Facial VII and Acusticus VIII	Abducens VI
4. 1st Branchial	1st Branchial	Glossopharyngeus IX	Absent
5. 2nd Branchial	2nd Branchial	Vagus X plus	Absent
6. 3rd Branchial	3rd Branchial	Accessorius XI	Hypoglossus XII
7. 4th Branchial	4th Branchial		
8. 5th Branchial	5th Branchial		

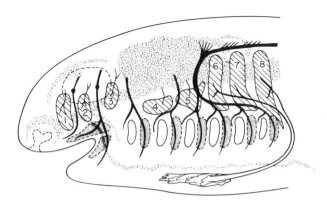

Fig. 3.27 Diagram showing the metameric origins of cranial organization.

iii The components of the dorsal root of the second or mandibular segment are the superficial ophthalmic, maxillary and mandibular rami of the trigeminal, supplying the snout, maxillary and mandibular regions respectively. The corresponding ventral root is the trochlear or patheticus (IV) which innervates the superior oblique. The dorsal root of the third or hyoid segment is the compound facial nerve (VII) together with the acoustic (VIII). The facial has the following branches:

> superficial ophthalmic—running forwards over the eye in company with the trigeminal and innervating the supraorbital lateral line organs;
> buccal—running forwards beneath the eye and innervating the infraorbital lateral line organs;
> hyomandibular—passing down behind the spiracle to innervate the lateral line organs of the lower jaw and the muscles of the hyoid arch;
> palatine—innervating taste buds on the roof of the mouth;
> pretrematic—passing down in front of the spiracle to innervate sense organs.

The acoustic nerve which innervates the sense organs of the ear is really an enlarged and specialized ramus of the facial. The corresponding ventral root is the abducens (VI) which supplies the posterior rectus muscle of the eye.

iv The dorsal root of the fourth segment is the glossopharyngeal (IX). It has a branch to the temporal region of the lateral line, a pharyngeal branch to the gut, and a branchial branch which divides to send a small pretrematic ramus to the hyoid arch in front of the first gill slit, and a main post-trematic ramus behind the first gill slit. The glossopharyngeal, therefore, bears the same relationships to the first gill slit as the facial does to the spiracle. There is no ventral root to the fourth segment; the somite which it would innervate disappears, and the fifth also lacks one for analogous reasons.

v Each of the remaining gill slits, 2–5, have branchial nerves with pretrematic and main post-trematic rami. These are actually the dorsal roots of segments 5–8 inclusive which unite to form a single compound nerve, the vagus, which also has two further rami arising outside its segments of origin. The lateral line branch reaches to the caudal extremity and the parasympathetic visceral branch supplies the heart, stomach and small intestine. These presumably permit the coordination of diverse visceral functions. The ventral roots of the sixth and following segments contribute to the hypoglossal nerve and innervate the relevant myotomal derivatives which contribute to the hypoglossal musculature. The nerve passes back over the gill slits, down behind them, and forwards beneath them. The ninth segment is the first to have a fully formed mixed nerve. It is therefore clear that despite its complex, derivative organization, the cranial region of gnathostomes retains the arrangement of separate dorsal and ventral nerves that characterizes the entire spinal cord in cyclostomes.

(b) The spinal cord and nerves

i Behind the cranial region the spinal cord has a entral and dorsal root on each side in each segment. The dorsal root bears a ganglion and the two roots unite to form a mixed spinal nerve, their components subsequently diverging to reach their individual destinations. In broad terms, the dorsal roots carry incoming information from the sense organs but also include some fine unmyelinated motor fibres. The cell bodies of the afferent neurons are situated within the dorsal root ganglia and after entering the spinal cord the afferent fibres terminate and make synaptic connections with other neurons. The ventral roots distribute motor fibres to the myotomal derivatives of trunk and fins, and, in addition, contribute a branch to the sympathetic system via the rami communicantes. These preganglionic fibres go to the sympathetic ganglia which form an irregular series, approximately metamerically arranged, lying dorsal to the posterior cardinal sinus and closely associated with chromaffin tissue with which they share a common origin from the embryonic neural crest. The postganglionic fibres go to the smooth muscles of the viscera.

ii Within the spinal cord the cell bodies of neurons are in, and indeed comprise, the centrally disposed grey matter. The axons make up the surrounding white matter. The grey matter can be separated into four longitudinal columns on each side. The most dorsal strip is where the axons of the somatic afferent neurons terminate. Beneath this lies the area of termination of visceral afferent axons. Under this again lie the cell bodies of efferent visceral neurons and, finally, the most ventral grey contains the cell bodies of somatic efferent neurons. This basic organization also underlies the structure of the medulla oblongata (q.v.) and the four regions are known respectively as the spinal somatosensory, spinal viscerosensory, spinal visceromotor and spinal somatomotor columns.

iii Many afferent fibres synapse with interneurons, a single isolated spinal segment can mediate both homo-lateral and contralateral responses, and there are well marked spinomedullary and spinomesencephalic tracts. Stimuli which augment the locomotory rhythm in intact fish have the opposite effect in gross spinal preparations. When free from contact, these preparations exhibit an overt locomotory rhythm of about 40 body undulations per minute. Gentle tactile stimulation, particularly of the ventral surface, can either increase or inhibit this. Rhythmic movements of both body and fins are abolished by complete deafferentation, but following local deafferentation the rhythm can be transmitted down the body by peripheral connections.

(c) The Brain

i THE MEDULLA OBLONGATA

The brain of *Squalus* is depicted in Fig. 3.28. The medulla oblongata lies between the spinal cord and mesencephalon and it grades into the cord posteriorly. During development, rostro-caudal waves of cell proliferation become superimposed upon an earlier, transverse, neuromeral organization. These give rise to longitudinal cell columns in the walls of the fourth ventricle. The roof remains membranous and gives the choroid roof. In selachians, longitudinal columns disposed in a dorso-lateral to ventromedial array are demarcated by regularly arranged depressions in the ventricular wall (Fig. 3.29). They correspond to the spinal columns mentioned above, the somatosensory and viscerosensory components forming the so-called alar plate, and the visceromotor and somatomotor columns forming the basal plate. The somatosensory

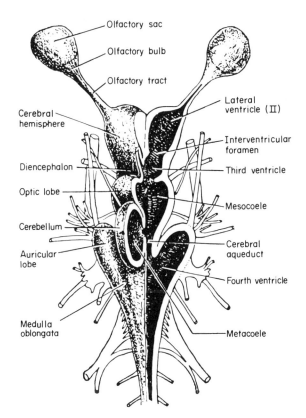

Fig. 3.28 Dorsal view of the brain of *Squalus acanthias* with part of the wall removed to show cavities.

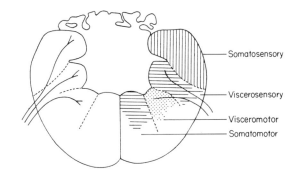

Fig. 3.29 Diagram of a selachian medulla showing longitudinal columns.

zone can be further divided into a dorsal region associated with the lateral line system and a more ventral one involved in more generalized somatosensory functions. The input from lateral line canals, cephalic sensory canals and ampullary organs passes to a huge 'balance' centre which may be differentiated into dorsal and medial vestibular nuclei or form a covering over all other medullary structures. A relatively straightforward situation characterizes *Hexanchus*, whose medulla resembles that of embryonic *Squalus*, but in *Squalus* and *Raia* a subsequent folding process complicates the picture.

Nuclear moieties associated with VII, IX and X occur in the viscerosensory zone with X making the paramount contribution. The visceromotor column encloses motor nuclei of V, VII, IX and X, and the somatomotor column includes nuclear aggregations of VI and XII. All occur in more or less longitudinal arrays from the front backwards.

The medulla also includes large multipolar cells of the reticular formation. As in cyclostomes, the fibres of these large reticular cells bridge the gap between anterior sites of stimulation and the motor apparatus of trunk and tail. Conduction velocities reaching $60\,\mathrm{m\,s^{-1}}$ are amongst the fastest recorded in poikilotherms.

ii THE CEREBELLUM

The histological structure of the cerebellum is constant throughout the gnathostomes. An external molecular layer is separated from an internal granular layer by an intervening layer of giant Purkinje cells. The enunciation of the 'dynamic loop' hypothesis for cerebellar function (see page 150) is, therefore, assumed to be of general application although needing appropriate modifications for the precise fibre relationships in a given taxon. In general terms the cerebellum receives proprioceptive impulses from the muscles, via the spinal cord, and sensory input from all exteroceptors and interoceptors. It coordinates muscular movements, maintains equilibrium, and its gross size reflects both body size and relative activity. In general, the greater the overall body size the greater the degree of overt lobulation. The median longitudinal fissure, reflecting its bipartite origin from dorsolateral upgrowths from the hind brain, is always present, but the transverse fissures vary. They are absent from species of small size such as *Spinax*, and also from *Scymnus*, *Hexanchus* and *Heptanchus*. In contrast they are deep in *Lamna*. The resulting anterior and posterior components are of equal size in Torpedidae, the anterior lobules are the smaller ones in Raiidae, the larger ones in Trygonidae and Myliobatidae. Lesioned preparations demonstrate the existence of fibre projections to the extrinsic eye muscles, to cell groups in the reticular formation, and to the thalamus.

iii THE MESENCEPHALON

The basic pattern of the midbrain in all other vertebrate classes is not unduly dissimilar from that of cyclostomes. In fishes a trigeminal nucleus is situated in the deep layers of the optic tectum and the nuclear aggregations of the trochlear and oculomotor nerves always have a mesencephalic location. The tectum comprises prominent bulges on the outside, is less complex than that of teleosts, and is partially overlain by cerebellar components. The more ventral tegmentum includes relay stations analogous to those in other taxa.

iv THE DIENCEPHALON

The diencephalon is detectable as a depressed region between the cerebrum and optic lobes when the brain is viewed from above. Most of its roof is extremely thin, forms the anterior tela choroidea and vascular folds of the choroid plexus hang down from it into the third ventricle. A transverse arched fold of tissue lying underneath the anterior ends of the optic tectum is the habenula—the hind end of the epithalamus. The thalamus is an important site of relays on both ascending and descending pathways and the hypothalamus is a centre for autonomic integration and the regulation of adenohypophysial functions. The first-named receives retinal input as well as spinal, cerebellar and tectal afferents. The last-named can be very complex with extensive preoptic nuclei that contribute fibres to the preopticohypophysial tract (c.f. hypophysis above).

v THE HEMISPHERES

As with all vertebrate brains, apart from those of bony fishes, the telencephalon arises during development as an unpaired median element that is continuous with the diencephalon behind. The two together form the prosencephalon and the primary telencephalon has been tentatively homologized with the archencephalon of Amphioxus. Its anterior limits are defined by the lamina terminalis. Subsequently it becomes separated from the diencephalon by a transverse epithelial fold, the transverse velum, and by the preoptic recess. Paired lateral evaginations from this median structure then give rise to the cerebral hemispheres. The gradual elaboration of these structures provides distinct dorsal and ventral (basal) hemisphere regions in which the emergent adult structures are represented by nuclear masses, fibre bundles and eminences in the walls of the lateral ventricles. Unfortunately, the lack of any universal agreement about the homologies of hemisphere structures in different classes has led, in the past, to a very diverse nomenclature and the literature is bedevilled by a host of terms whose differing usage by various workers has led to a tangle of complete and partial synonymies.

If one inserts a caveat about the frequent lack of distinct boundaries between different nuclear masses, it is possible to summarize the overall organization. A septal region is broadly transitional to the diencephalon and associated with dorsal pallial structures and deeper striatal aggregations. The hippocampus, otherwise known as the archicortex or archipallium, is the predominant dorsomedial hemisphere component. The pyriform lobe, paleocortex or paleopallium is situated laterally. Between them lies the general pallium. Beneath these are hyperstriatal, neostriatal and paleostriatal components. These are indistinct and controversial in selachians and holocephalans. In the past, the principal fibre relationships of selachian forebrains have been attributed to olfactory associations but in *Ginglymostoma, Galeocerdo* and *Negaprion* these are now known to represent a relatively modest component.

FOSSIL FORMS

i Although cartilage does not fossilize as easily as bone, many fossil genera have long been known. There are three major radiations—the cladodont, hybodont and modern. The best preserved of the early ones originate from the late Devonian Cleveland shales. Nodules containing remains of *Cladoselache* provided teeth, calcified skeletal cartilages, impressions of the skin and outlines of the body (Fig. 3.30). These demonstrated that there were two dorsal fins with spines in front of them (cf. *Squalus*) and pterygiophores reaching

almost to their edges instead of having the bulk of the fin supported by ceratotrichia as in modern sharks—the pterygiophores of the Raiiformes approximate to the pectoral fin margins. The tail fin was markedly heterocercal and there were well-developed pectoral and pelvic fins, the former being much the larger. The jaws were longer than those of modern sharks, and the mouth reached the tip of the snout, but were not unduly dissimilar to those of *Chlamydoselachus*. The jaws had an amphistylic suspension recalling that in *Cestracion, Hexanchus* and *Heptanchus*. The hyomandibular articulated with a prominent postorbital process and probably with an additional prominence further forward at the base of the braincase.

In *Ctenacanthus*, present from the Upper Devonian to the Permian and common in the Carboniferous, stout spines with numerous beaded longitudinal ridges acted as cutwaters for the dorsal fins. The paired fins had multiple radial elements, rather than the single rods of *Cladoselache*, and it is probable that there was a normal anal fin unlike the double structure of *Pleuracanthus*. Predaceous cladoselachians were generally widespread in the Carboniferous but became extinct during the Permian. Various other forms replaced them in the Mesozoic Era, many of them being so-called hybodonts. In some of these genera, e.g. *Helicoprion*, a central series of teeth were presumably situated in the jaw symphysis and formed an intriguing whorl.

ii Fossil forms reminiscent of rabbit fishes were mentioned in the introduction to this chapter in the context of *Ctenurella*. They have also, in the past, been suggested as derivatives from the bradyodont stock. This included a variety of forms and, in particular, cochliodonts were

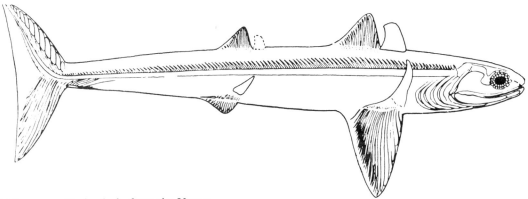

Fig. 3.30 The genus *Cladoselache* from the Upper Devonian. (After Harris.)

seen as possible intermediates between selachians and chimaeras. There are, indeed, various similarities between fossil genera known as bradyodont and holocephalans. The skulls of *Helodus, Chondrenchelys* and *Fadenia* have a well-defined ridge in the anterior part of the otic region connecting a postorbital process with the subotic shelf. In *Helodus*, as in chimaeras, there were large pectoral fins and the upper jaw was solidly fused to the braincase. However, such similarities could be the result of convergent evolution.

MODERN HOLOCEPHALIANS.

The chimaeras are essentially benthic and occur at depths of 100 to 1500 metres. There is a marked sexual dimorphism with the females attaining a considerably larger size than the males. The spine in front of the first dorsal fin is venomous. Eggs are laid in pairs and have a horned capsule with a characteristic shape.

FURTHER READING

ALEXANDER R. McN. (1967) *Functional design in fishes.* Hutchinson University Library, London.

GILBERT P.W., MATHEWSON R.F. & RALL D.P. (1967) *Sharks, skates and rays.* Johns Hopkins Press, London.

GILBERT P.W. (1963) *Sharks and survival.* D.C. Heath & Co, Boston.

HOAR W.S. & RANDALL D.J. (eds) (1969) *Fish Physiology*, 7 vols. Academic Press, London.

KLEEREKOPER H. (1969) *Olfaction in fishes.* Indiana University Press, Bloomington and London.

4 · Class Osteichthyes 1. Actinopterygii

Synopsis

Subclass Actinopterygii

Superorder Chondrostei —*Paleoniscus
                          Acipenser
Superorder Holostei —Amia
Superorder Teleostei —Salmo

INTRODUCTION

The remaining fishes all have a bony skeleton—are Osteichthyes. The Actinopterygii are the ray-finned and spiny-finned fishes such as the eels, carp and flying fish whose paired fins are broadly fan-shaped. Most authorities attribute living genera to three infra-classes or superorders. *Polypterus, Calamoichthys, Acipenser* and *Polyodon* (Fig. 4.1) are placed in the Chondrostei, to which the fossil paleoniscoids are allied. The two other relict genera *Lepisosteus* and *Amia*, together with many fossil forms, are Holostei (Fig. 4.2). All other actinopterygians belong to the Teleostei. These animals have undergone an immense amount of speciation since their origin in the Mesozoic Era and various evolutionary trends are apparent.

INTEGUMENT

i The epidermis comprises layers of cells held together by viscous material and containing tactile receptors. During development it gives rise to taste buds, lateral line organs, light organs and poison glands. The many mucous glands protect the skin from bacterial and fungal infections as well as contributing to ionic regulation.

ii The overall body form, again closely associated with hydrodynamic properties, is maintained by the thick fibrous dermis from which the scales are derived and in which they are embedded. The skin must yield to the wave-like flexures of the body involved in locomotion and these vary from the undulations of eels to the barely perceptible bending of the tunny fish body. Such varying degrees of flexure are reflected in the fibre patterns of the dermis. The layering of the principal tracts of fibres varies but some always pursue left-handed, others right-handed, geodesic spirals. They thus form a lattice over the myotomes, crossing each other at angles which are smaller in deep-bodied fish than in slim ones (Fig. 4.3).

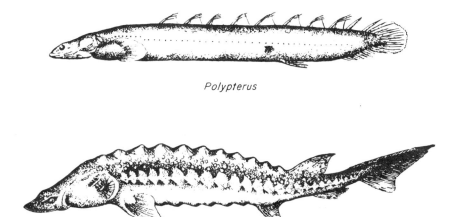

Polypterus

Acipenser

Fig. 4.1 The genera *Polypterus* and *Acipenser*.

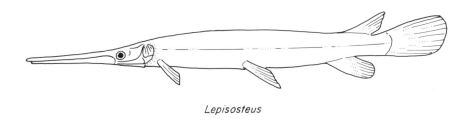

Lepisosteus

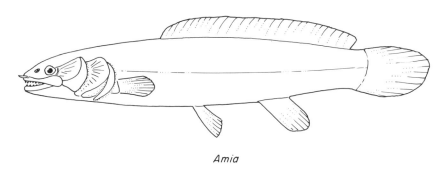

Amia

Fig. 4.2 The genera *Lepisosteus* and *Amia*.

iii Bony scales may have been derived from primary dermal armour (but see page 25), and those of primitive actinopterygians, like those of acanthodians and crossopterygians (q.v.), have the same composition as ostracoderm and placoderm armour. This comprises a deep layer of bone, a middle layer of dentine and an outer layer of enameloid. Paleoniscoid scales, and those of *Polypterus* and *Calamoichthys*, are deeply embedded, fit into each other by a characteristic peg and socket joint, and grow by addition to both the bony isopedin layer and the surface ganoin (Fig. 4.4). An intermediate pulp layer corresponds to the cosmine of crossopterygian scales.

Lepisosteoid scales characterize both fossil and living Holostei. The original ganoid scales have become thinner in successive genera along several phyletic lines. The scale is then formed from two layers—isopedin and ganoin (Fig. 4.5). *Amia*, with a rudimentary layer of ganoin, suggests the origin of teleost scales. Amongst these a reduction of the ganoin and isopedin layers gives elasmoid scales of varying form. They can have a uniform appearance or be divided into anterior, posterior and lateral fields, of which only the posterior is visible during life. Concentric rings mark the annual growth and are widely used for ageing the fish. Two principal types occur. Cycloid scales are the first to appear in the fossil record and today typify salmonids, clupeids, cyprinids, characids, etc. They are devoid of spinules. Genera that are considered to have diverged further from the ancestral stock, for example percoids, scombroids and soleoids, have ctenoid scales in which the rear free edge is toothed.

As scales develop from the dermis their disposition follows the geodesic spirals of the dermal fibres. In the tarpon one row spirals round in a continuous course from head to tail whilst an intersecting row courses in the reverse direction. The pattern of stresses in a moving fish is thus reflected in the structural patterns of both dermal fibres and scales. Scale size is also affected. In the herring or carp, fishes with fairly large scales, the most supple part of the body, the tail, is invested with smaller scales than the less mobile trunk.

iv Camouflage is a further integumentary attribute. Bottom-living genera have colours which merge with the substratum and are broken up into irregular patterns

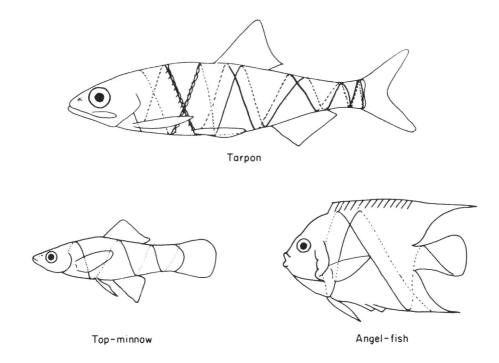

Tarpon

Top-minnow

Angel-fish

Fig. 4.3 The scale rows of fishes showing their organization on geodesic spirals. (After Breder & Marshall.)

that help to render the fishes inconspicuous. Forms that habitually swim in mid-water must merge with the bright background of surface water when viewed from below, and the dark background of bottom water when viewed from above. Although they often have an overall silvery appearance, counter-shading is widespread with the dorsal surfaces appropriately dark. Complementarily, the ventral surface is frequently narrow thereby minimizing the conspicuousness of the ventral strip which cannot be camouflaged when viewed from below. The silveriness is a property of reflecting platelets that comprise crystals of guanine with intervening layers of cytoplasm and are either attached to the inner surface of scales or embedded in the dermis. Individual platelets reflect particular colours best but superimposed layers of platelets enable the fish to reflect effectively through the entire spectrum.

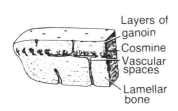

Fig. 4.4 A diagram of a ganoid scale.

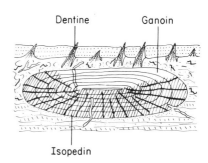

Fig. 4.5 A scale of *Lepisosteus*.

SKELETAL SYSTEM

(a) The skull

i During ontogeny a cartilaginous skeleton, homologous with that of Chondrichthyes, is replaced by bone. The cranial skeleton has a dual origin during phylogeny. Dermal plates, comparable to large scales, and internal bones, replacing embryonic cartilages, provide an ossified braincase, a shield of dermal bones covering the top and sides of the head, an upper jaw and palatal structures. It is impossible to be sure of the homologies in different groups but the pattern in the paleoniscoid genus *Cheirolepis* is shown in Fig. 4.6. It comprised:
> a paired series of bones on either side of the mid-dorsal line involving nasals, frontals, parietals and postparietals;
> marginal tooth-bearing maxillae and premaxillae;
> a circumorbital ring including prefrontal, postfrontal, intertemporal, postorbital and jugal;
> lateral elements of the posterior roof comprising supratemporal, tabular and extrascapulars.

ii Jurassic genera within the holostean lineage had additional components. In teleost genera such as *Gadus*, the cod, the floor of the braincase consists of basioccipital, prevomer and parasphenoid; the meseth-moid is anterior. Nasals, frontals, parietals and a supraoccipital form the roof. The auditory capsules are well ossified with prootic and opisthotic bones below; sphenotic, pterotic and epiotic bones above. The sides of the braincase are very incomplete. Anteriorly there are paired prefrontals, further back there are the paired laterosphenoids. A string of small bones bounds the orbit—lachrymals, suborbitals and post-orbitals (Fig. 4.7).

(b) Jaws and branchial arches

In the fossil paleoniscoids the mouth was elongated and the jaw was supported by the hyomandibular, or sometimes an intermediate bone of the hyoid arch, the symplectic, and by ligaments binding its anterior end to the cranium. This type of suspension has been called methyostylic. In the Triassic sub-holostean genera the jaw articulation was further forward. In holosteans the hyomandibular, quadrate and epipterygoid bones are all enlarged as jaw supports and the symplectic may unite the first two. The maxilla is relatively reduced. This presages the teleost condition where the lengthened premaxilla can cut the maxilla off from the edge of the mouth. In advanced teleosts it then contributes to the protrusion of the mouth by pushing the premaxilla forward as the mouth opens.

The lower jaws consist of cartilage and cartilage bone,

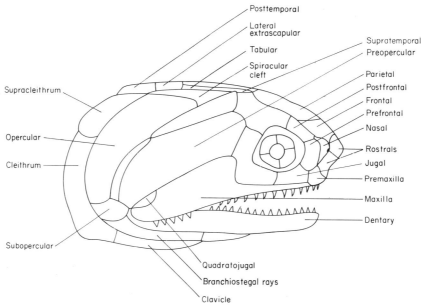

Fig. 4.6 The skull of *Cheirolepis*.

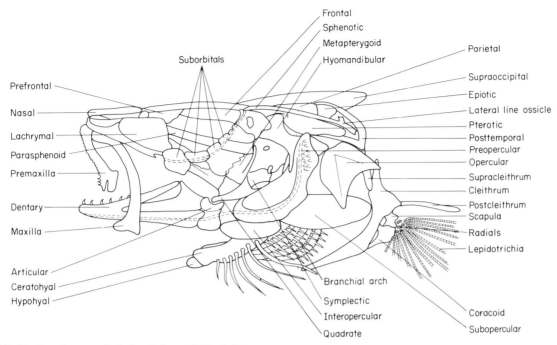

Labels in figure:

Suborbitals · Frontal · Sphenotic · Metapterygoid · Hyomandibular · Parietal · Supraoccipital · Epiotic · Lateral line ossicle · Pterotic · Posttemporal · Preopercular · Opercular · Supracleithrum · Cleithrum · Postcleithrum · Scapula · Radials · Lepidotrichia

Prefrontal · Nasal · Lachrymal · Parasphenoid · Premaxilla · Dentary · Maxilla · Articular · Ceratohyal · Hypohyal

Branchial arch · Symplectic · Interopercular · Quadrate · Coracoid · Subopercular

Fig. 4.7 Skull and pectoral girdle of the cod (*Gadus*) to show complexity.

homologous with Meckel's cartilage, invested by dermal bone. The hyoid bars and gill supports are ossified equivalents of those in selachians. The opercula, and the branchiostegal rays below them, consist of dermal bone.

(c) Axial skeleton

i The axial skeleton of fossils is frequently obscured by the thick scales. No centra occur in Chondrostei where the notochord is persistent, unconstricted and surrounded by a thick fibrous sheath. *Amia* has a well-ossified column. In the anterior trunk region amphicoelous (biconcave) centra have paired ventrolateral processes, or parapophyses, bearing pleural ribs. Separately ossified neural arches bearing paired spines rest on the postero-dorsal surface of one centrum and the antero-dorsal surface of the next. Further back they rest entirely on one centrum and the spines are fused into a single structure. In the anterior caudal region a pair of separately ossified haemal arches bearing a haemal spine lie below each pair of neural arches. From the sixth or seventh caudal segment backwards two similar biconcave centra occur in each segment, only one of which bears haemal arches. In the posterior upturned region

(see below) the haemal arches fuse with their centra. Throughout the column dorsal and ventral spinal nerve roots emerge between successive pairs of neural arches. Similar neural and haemal arches are usually well-formed in other Actinopterygii but, as in the case of Amphibia (c.f. page 75), it is probable that they all have slightly different embryological origins from those in selachians.

ii The vertebrae of *Lepisosteus* are unique amongst fishes as they are opisthocoelous, the centrum having a convex anterior face which fits into a concavity in the rear of the one in front. Teleosts usually have well-developed amphicoelous bony centra between which the notochord remains thick. In general they show no trace of a compound origin but develop as a cylinder. Ossification may then unite centrum, basidorsals and basiventrals into a single structure but the basidorsals remain independent in clupeids and cyprinids, the basiventrals in *Esox*, the pike. Special intervertebral articulations can be formed from either neural or haemal arches or from the centra themselves.

iii Paleoniscoids had heterocercal tails and these persist in Chondrostei. The buoyancy of teleosts is maintained by their swimbladders (page 56), heterocercal tails are not necessary and the hind end of the vertebral column is modified to form the base of a symmetrical or homocercal tail. This is foreshadowed in holosteans. In some primitive teleosts, such as the order Clupeiformes to which the herring belongs, the notochord is turned upwards and enclosed in a urostyle representing fused centra. This bears the enlarged ventral hypochordal fin-lobe. The most characteristic modification is, however, the development of hypural bones. These are numerous in clupeomorph fishes but reduced in number, increased in size, and disposed symmetrically above and below the longitudinal axis in most specialized forms (Fig. 4.8).

iv One group of teleosts, the Ostariophysi, have anterior vertebrae modified to form Weberian ossicles which confer increased sensitivity to sound (q.v.). The first two ossicles are modified neural arches, the third is the foreshortened rib of the third vertebra. An early stage in their evolution is represented in the tropical order Gonorhynchiformes.

(d) Pectoral and pelvic girdles

i The pectoral girdle always includes scapulo-coracoid components analogous to those of selachians. Large in Chondrostei and Holostei, they remain cartilaginous in *Amia* but the scapular region ossifies in *Lepisosteus*. Small scapular and coracoid ossifications also occur in teleosts. Additional dermal bones vary in prominence. Large clavicles occur in paleoniscoids and *Polypterus*, with large cleithra and smaller postcleithra, supra-cleithra and post-temporals above them. The clavicles

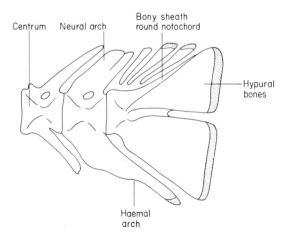

Fig. 4.8 Skeleton of caudal region in a goby. (After Lotz.)

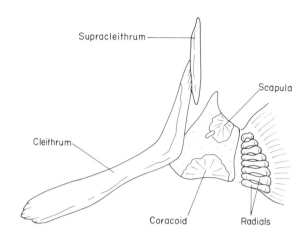

Fig. 4.9 The pectoral girdle of an eel, *Anguilla*, seen from the left side.

are absent from other actinopterygians where they are replaced by forward extensions of the cleithra (Fig. 4.9).

ii The two halves of the pelvic girdle, ossified in all but *Acipenser* and *Polydon*, form two separate bones lying horizontally in the body wall and apposed in front of the anus to give a Y-shaped effect (Fig. 4.10). Small additional cartilages may occur anteriorly and the plates of the two sides may meet in a median cartilage as in *Gadus*. The fin skeleton is borne at their divergent posterior ends.

(e) Pterygiophores and lepidotrichia

i In paleoniscoids the greater part of the fins was supported by dermal fin rays. Although the paired fins were broadly based and triangular in early forms there was, at most, only a small muscular basal lobe. In *Polypterus*, two bones, the pre- and post-axial radials, flank the proximal region of the pectoral fins and the intervening volume is cartilage. In the pelvic fins of *Polyodon*, serially repeated radials articulate with processes on the pelvic girdle.

ii The pterygiophores of teleosts comprise cartilage bone and the rays are bony, jointed and branched lepidotrichia that contrast with the horny, unjointed ceratotrichia of selachians. Each consists of two half-rays, one on either side of the fin, and articulates with distal pterygiophores. These are made up of two halves bound together by collagen. The base of the ray straddles the apex of the pterygiophore row and, attached by ligaments, forms a loose joint enabling the

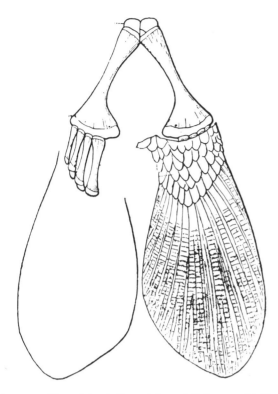

Fig. 4.10 The pelvic girdle and fins of *Polypterus bichir*.

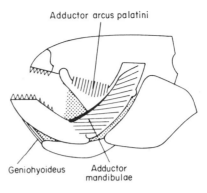

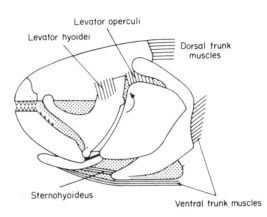

Fig. 4.11 Diagrams of some important muscles in the head region. (After Alexander.)

fin to be opened or closed in a fan-like manner. Lateral movements of the rays are facilitated by a longitudinally oriented hinge joint between the ultimate and penultimate pterygiophores (cf. muscles, below).

iii Many fish such as perch and sticklebacks have spines. In catfish these are formed by the anterior rays of both pectoral and dorsal fins. The component lepidotrichia are fused into a unified structure and, when erect, they render the fishes difficult to swallow. In many teleosts (e.g. *Trachinus*, the weevers) they are associated with poison glands.

MUSCULAR SYSTEM

i Much of the cephalic musculature is involved in opening and closing the mouth during feeding. The combined actions of the sternohyoideus and trunk musculature displaces the hyoid bars laterally, widens the head, and prevents the tongue moving forward. During mouth closure, the geniohyoideus returns the hyoid bars to their original position whilst the adductor

arcus palatini returns the sides of the buccal cavity and the operculum to theirs (Fig. 4.11).

ii Trunk myomeres are comparable with those of selachians. Epaxial units are inserted anteriorly on to the skull and pectoral girdle, whilst hypaxial units below the lateral line form the abdominal wall. Separate longitudinal muscles, the supracarinals and infracarinals, run the length of the body at the extreme dorsal and ventral margins. The dorsal ones are usually interrupted at each dorsal fin but in some eels which have a continuous fin they continue uninterruptedly and have separate insertions on each fin element. Their principal function is to produce flexures of the dorsal and ventral regions.

iii Movements of the dorsal fins are accomplished by levator and depressor muscles, which raise or lower the rays, and by inclinator muscles that displace the fins

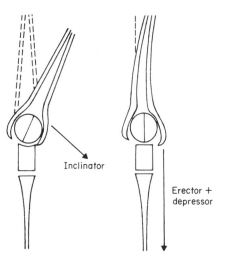

Inclinator

Erector + depressor

Fig. 4.12 The muscles acting on the rays of a fin. (After Alexander.)

laterally about the longitudinal joint (Fig. 4.12). The caudal fin musculature varies. Comparatively simple extensions of the caudal myomeres are inserted on the fin ray bases in *Polypterus*, but in teleosts various flexors and adductors originate on the posterior vertebrae. The pectoral fins, whose precision of movement can be crucial during manoeuvring, have abductors, originating on the coracoid, that depress them and close their rays. Extensors, originating on the cleithrum and inserted on the outer surface of the first ray, spread the fins horizontally. Superficial and deep adductors, originating on the coraco-cleithral junction, draw them against the body. Interfilamentous units between the rays close them. The infracarinals contribute to the movements of the pelvics but these also have adductors and abductors that are analogous to those of the pectorals.

iv Slow and fast fibres were first characterized in teleosts. As usual they differ in diameter, mitochondrial and lipid content. In many genera such as catfish and mollies the fibres have ribbon-like myofibrillar bundles around their edges. Large diameter fast fibres, equivalent to the type II fast fibres of selachians, can be surrounded by a regular array of smaller fibres (type I of selachians) giving a mosaic arrangement. Electromyographic studies on salmon show that superficial slow fibres are active at all speeds but that mosaic muscle is inactive below 75 % of the maximum cruising speed. At these speeds the fast fibres exhibit electrical activity similar to that in slow fibres, but at very high speeds much larger electrical events occur.

ELECTRIC ORGANS

Electric organs occur in at least three unrelated taxa. The electric eel, *Electrophorus electricus*, has huge structures representing 40 % of the body volume and producing pulses of 500–600 V for defence or stunning prey. It also has a smaller organ that produces regular 10 V pulses. As in other gymnotoids, and the unrelated mormyrids, this is used in conjunction with electroceptors (q.v.) both to locate conductors and insulators in the environment and to provide social signals. The organs, which can be modified muscle or nerve fibres, form some four rows of drum-shaped electrocytes along the body. Action potentials occur across both the anterior and posterior surfaces but those across the latter start more quickly and end sooner than those across the former. The rostral side of the cell is, therefore, successively positive, then negative, to the caudal side. The actual potentials measure 100–150 mV but as those on each surface overlap in time the net potential across the cell is far less—c. 50 mV. The discharge frequency varies from a few per minute to c. 600 Hz, or 1700 Hz for those of neural origin.

LOCOMOTION

i The presence of swimbladders (q.v.) releases the pectoral fins of teleosts from an inordinate involvement in controlling the depth at which the animals swim. This is associated with great diversity in their form and function. During fast swimming, typical teleosts keep all their fins, other than the caudal one, folded flat. They are only brought into use to counteract any forces tending to displace the animal either upwards or downwards, for braking, and for precisely controlled locomotion. The swimbladder confers upon the fish a density equivalent to that of the surrounding water at a particular level. If the fish is displaced upwards, the pressure on the swimbladder will be reduced, and the swimbladder will, therefore, expand, further reducing the fish's density, and the animal will continue to rise. Complementarily, a downward displacement will contract the swimbladder, thereby increasing the density of the fish, and it will continue to sink. The paired fins are used to correct such passive displacements when they are small and undesirable. However, when the fish requires to move up or down in the water the swimbladder must have its pressure modified to adjust to the change in external pressure experienced. How this is done is explained later.

Spreading the pectoral fins at right angles to the

direction of motion provides a very effective braking action. However, if they were situated low on the body the braking force would act below the centre of gravity and tilt the head downwards. A position in the horizontal plane which passes through the centre of gravity avoids this tilting action. In acanthopterygians (q.v.) the pelvic fins occupy the position vacated by the pectorals and the two sets act synergistically to prevent tilting. A downward force on the pelvics counteracts an upward lift on the pectorals.

When stalking prey a pike glides through the water using its dorsal and pectoral fins. At a faster pace the other fins come into use. It is only during the final raptorial dart that full body flexures are involved. Eels, halosaurs and notacanths, which are elopomorph teleosts with long, tapering bodies often lacking caudal fins, habitually utilize undulations along the dorsal or anal fins, or quick strokes of the pectorals. Additional thrusts for speed are contributed by waves of flexure along the body. Similar effects occur in the Mormyridae and Gymnotoidei, which are representatives of the unrelated superorders Osteoglossomorpha and Ostariophysi.

ii Amongst the superorder Acanthopterygii some members of the order Gasterosteiformes, to which the stickleback belongs, have their body transformed so that gross flexures are not possible. Thus the snipe fishes, *Macrorhamphosus*, and the bellows fishes, *Centriscops*, have their skin armoured with rough scales and bony scutes. Undulations of the dorsal and anal fins are then the sole sources of propulsion. Similarly, shrimp fishes, Centriscidae, which hover head downwards, have immovable plates around their compressed body which itself ends in a dorsal spiny process rather than the caudal fin which is displaced downwards. The short tail, directed downwards and backwards from the ventral side, together with the dorsal and anal fins propel the body. In the related Syngnathidae, the pipe fishes and sea horses, tail fins can be absent. High speed cinematography reveals that the individual fin rays of the dorsal and pectoral fins of sea horses oscillate from side to side at up to 70 Hz. If the undulations are of small amplitude, the thrust produced is parallel with the base of the tail. When the fin rays are further displaced the thrust is away from the fin base.

iii Locomotion in air and on land remain to be discussed. Flying fish leave the water to evade predators. In the genus *Exocoetus* (superorder Atherinomorpha) the extensive pectorals with their strong flexible fin rays are located high up on the body. In *Cypsilurus* and

Prognichthys the pelvics are also wing-like. Impelled by vigorous, rapid strokes of both the tail and the caudal fin, whose lower lobe is enlarged, they travel through the water at speed. Inclined at $15°$ to the horizontal they leave the surface at 15–20 mph with the pectorals open and the lower lobe of the tail fin still beating at 50 Hz in the water. The pelvics are then unfolded, generate lift, and bring the fish to the horizontal. By closing the pelvic fins and lowering the tail fin into the water they can regain speed for 2–5 consecutive renewed glides. Some related hemirhamphids with large pectorals can perform similarly. Flight sustained by active strokes of the pectorals only occurs in freshwater fishes such as the South American hatchet fishes, *Gastropelecus*, and unrelated west African butterfly fishes, *Pantodon*.

iv Lizard fishes, Synodontidae, and gurnards, *Trigla*, can creep along the sea-floor using their pelvic fins as struts. Sargassum-fishes, *Histrio*, clamber over sargassum using their fins as hooks and levers. A number of both freshwater and marine species can also move on land. Serpentine forms like the eels move on land by waves of flexure passing down their bodies. The Indian climbing perch, *Anabas*, has sharp spines on its gill covers which are spread and fixed on the ground whilst the body is pushed forward by the pectoral fins and tail. The snakeheads, Ophiocephalidae, use their pectorals as levers. In *Periophthalmus*, the mud skippers, the pectorals are hinged to the pectoral girdle and strong, lower rays are bunched together forming struts. The weight of the body is borne by the pelvics and tail. *Malthe vespertilio*, the bat-fish, uses both sets of paired fins to crawl in a manner that is reminiscent of frogs and toads.

v Finally it is worth mentioning some specialized uses of fins. The shark suckers, Echeneidae, have the first dorsal fin transformed into a sucking disc. In the lumpsuckers, Cyclopteridae, the fused pelvic fins form a similar structure by which they cling tenaciously to rocks. In the gurnards the anterior rays of the pectorals are modified to form feelers bearing chemoceptors. The first spine of the dorsal fin forms the angler's rod and lure, and the anal fin is modified to form an intromittent organ in all cyprinodontids.

DIGESTIVE SYSTEM

(a) Buccal cavity and feeding

Evolutionary trends within the Actinopterygii involve the mouth becoming terminal and reduced in length.

The tongue is a thickening of the floor of the buccal chamber and teeth are variously borne on premaxillae, maxillae, pterygoid, parasphenoid and glossohyal bones, with additional pharyngeal teeth on pharyngo-branchial IV and ceratobranchial V. Cyprinoids lack teeth in their jaws but have large lower pharyngeal teeth which press food against a dorsal horny pad.

In paleoniscoids the maxillae were fixed, in holosteans the hind end is freed from the side of the skull and in teleosts it can be free throughout most of its length. The premaxillae can also be free and both then lie in the membrane forming the side of the mouth. In holosteans, and those teleosts in which the premaxillae are fixed, the maxillae are hinged to the cranium anteriorly and attached to the lower jaw in front of the jaw articulation. When the jaw opens they swing forward, thereby enlarging the gape. In forms such as the herring both the maxillae and premaxillae are so hinged and both swing forward. In Cypriniformes the premaxillae are actually protrusible and move forward away from the maxillae. Various Acanthopterygii achieve a similar protrusion in different ways and its widespread occurrence implies that the resulting versatility of the lips is of great selective advantage when catching and manipulating food.

(b) Stomach and intestine

In some Chondrostei and some Holostei the oesophagus leads to a U-shaped stomach analogous to that of selachians. In some teleosts the ascending pyloric limb is very much reduced in length, whilst in *Polypterus*, *Amia*, and *Anguilla* (the eel), amongst teleosts, the descending cardiac limb is prolonged into a long caecum (Fig. 4.13). In many teleosts a distinct stomach is lacking. In

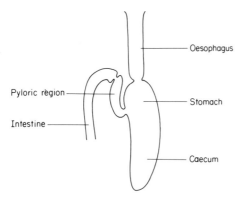

Fig. 4.13 Diagram of an eel stomach showing prominent blind caecal region.

contrast some clupeomorph and osteoglossomorph genera have the walls transformed to a muscular gizzard. From 1–1000 pyloric caeca can also be present. The intestine receives the bile duct, passes directly to the rectum and anus in Chondrostei and Holostei, but can follow a tortuous, and even coiled, course in teleosts.

SWIMBLADDER

i In *Polypterus* a predominantly ventral bilobed sac opens on the floor of the pharynx just behind the gill slits. Only the elongated right lobe attains a dorsal position in the mesentery above the gut. In the remainder of the Actinopterygii the airbladder is essentially a median dorsal diverticulum. In *Amia*, *Lepisosteus*, *Acipenser*, *Polyodon* and teleosts such as elopomorphs and pike it retains a pneumatic duct opening into the dorsal wall of the gut. This is the physostomous condition. Other teleosts lack the duct—the closed or physoclistous condition.

ii Teleosts which lack a swimbladder, such as the bottom-living flatfish, have a specific gravity of 1.06–1.09, comparable with that of those selachians which lack squalene. Those with swimbladders have specific gravities close to that of the water in which they are swimming. In shallow-water forms the gas mixture in the bladder approximates to that of the atmosphere. Those from deeper water have higher proportions of oxygen. Salmonid fishes, irrespective of their habitual depth, have high proportions of nitrogen. The partial pressures of gases in arterial blood are usually close to those of the water, however, the partial pressure of oxygen in the swimbladder is often somewhat greater than this so that oxygen tends to diffuse from it to the blood and its loss has to be compensated for.

iii Physostomous fishes renew the gases in the bladder by gulping air at the surface. Physoclistous genera have to achieve it entirely by secretion. Some genera have no specialized region for this purpose and direct observation shows that it occurs over much of the bladder wall. Others have definitive gas glands. Secretion involves the release of lactic acid into the blood. As an electrolyte this reduces the solubility of all gases in the blood and, by lowering the pH, initiates the dissociation of oxyhaemoglobin by the Bohr effect. The gas secreted into the swimbladder is, therefore, largely oxygen. Secretion at substantial depths relies on the precise arrangement of the blood vessels. The arterial supply forms long, parallel capillaries which mingle with venous capillaries leaving the region giving a complex rete mirabile. The venous blood has a higher partial pressure of oxygen due to the release of lactic acid, and oxygen tends to diffuse from the venous to the arterial side. This raises the arterial partial pressure above that of the arterial blood generally prior to its entering the gas gland. The partial pressure continues to rise until at the site of the gas gland it exceeds that in the swimbladder and oxygen is released into the latter. Clearly, the maximum partial pressure attainable is greater in long rete, and teleosts habitually living at great depths have longer rete than shallow-water species.

RESPIRATORY SYSTEM

i Gaseous exchange again predominantly involves the mouth and a one-way, constant flow of water over the gills. In teleosts such as the salmon the lateral wall of the buccal cavity has two articulations with the cranium which permit its lateral displacement, widening the cavity. The hind articulation is provided by the hyomandibular which is connected to the hyoid via the interhyal. The hyoid itself is also in contact with the operculum, to which it is attached by ligaments, and displaces the operculum laterally. The entire orobranchial cavity is thus enlarged and water is drawn in through the mouth. None can enter at the operculum because flexible tissue at its edge acts as a valve, and because the cavity is floored by a membrane stiffened by branchiostegal rays which are jointed to the hyoid bars and open in a fan-like manner when the head is widened. Subsequent narrowing of the head ejects water out past the hind edge of the operculum, or through the fused ventral opening in synbranchids. Valves prevent it from leaving at the mouth.

The gill lamellae resemble those of selachians, and gill-rakers, arising from the gill arches, restrict food to the alimentary canal. The surface area of the lamellae varies immensely and, for a given body weight, sluggish species have a smaller area than active ones. Thus an angler and tuna of similar weight had areas of 0.22 m^2 and 3.1 m^2 respectively.

ii Other organs can be involved in gaseous exchange. Various species inhabiting oxygen-poor water extract oxygen from the air taken in at the mouth. In *Polypterus* the air-bladder is a lung supplied with blood from the posterior branchial arches and drained by a vein entering the hepatic portal system. In *Erythrinus* the swimbladder is again involved and the middle one of its three chambers is honeycombed with highly vascular

walls supplied with blood from the coeliac artery and drained by the inter-renal vein. *Gymnotus* has highly vascularized papillae in its mouth which are supplied with blood from the branchial arches and return it to the heart by the cardinal and jugular veins. The stomach is a vascular, thin-walled respiratory organ in *Plecostomus*, the intestine in *Callichthyes*, and the rectum in cobitid loaches. *Clarias*, the African catfish, has capacious diverticula, with highly vascular arborescent processes, above the gill chamber on each side. They are supplied with blood by the second and fourth afferent branchials and drain to the efferent branchials. Highly vascular pharyngeal pouches occur in the Ophiocephalidae that emerge on land, and in the Anabantidae the gills are actually of reduced importance.

The swimbladder of *Anabas* has an anterior part walled by striped muscle and a posterior part that is divided into two longitudinal chambers—one on either side of the vertebral column. The fish comes to the surface and contraction of the somatic musculature expels air from the posterior chambers and into the anterior chamber. This expands and, in turn, expels air from the neighbouring labyrinth pouches. The body muscles then relax, the muscles of the anterior chamber contract and displace its contents into the posterior chambers, the opercula are raised and air streams into the labyrinth pouches. These are the site of gaseous exchange.

URINOGENITAL SYSTEMS

(a) Kidneys

In freshwater fishes the blood has an osmolar concentration of 130–170 mmol and there is a copious dilute urine. Various special devices are adopted to minimize the tendency to gain water and lose salts. The skin is not very vascularized and production of mucus improves the waterproofing. In the ocean the problem is to conserve water. The blood of marine teleosts is more concentrated, the urine is scanty and the kidney tubules are short. Indeed they can, as in *Opsanus tau*, the toadfish, totally lack glomeruli. Excretion by such aglomerular kidneys is essentially tubular secretion. Water and salts are taken in at the mouth and then excess salts are excreted by special chloride secretory cells in the gill.

(b) Gonads and reproduction

i The actinopterygians have various modifications of the basic system described for selachians. In *Acipenser*,

Lepisosteus and *Amia* vasa efferentia extend across to the mesonephros throughout the length of the testis. In *Polypterus* a sterile hind region of the testis consists of canals that lead to a single duct which opens into the urinogenital sinus. In the females the oviducts of *Polypterus* and *Amia* have wide funnels situated about halfway up the ovary. In *Lepisosteus* each ovary projects into a closed sac which opens into the base of the mesonephric duct. A few teleosts shed eggs from the ovaries into the general coelom, but in the majority the ovaries are enclosed in ovisacs which narrow behind to form oviducts.

ii Teleosts are not invariably male or female. Individuals of some sea-perch and sea-bream are functional hermaphrodites with one part of the gonad forming a testis, the other an ovary. Similar conditions occur in deep-sea families of tripod fish. Some groupers undergo a complete sex reversal from male to female as they grow.

iii Many forms of parental behaviour occur and the fact that fertilization is usually external often results in this being undertaken by the male, the female having no part in it. Nevertheless various degrees of ovo-viviparity and viviparity occur. Some top-minnows retain their eggs in the oviduct for a time and the guppy (a poeciliid) is well known as an ovo-viviparous species. Many poeciliids and anablepids are true live bearers. In such forms there is a single oviduct, usually a single ovary, the anterior rays of the male's anal fins form intromittent organs, and the young are nourished through some kind of 'placental' connection. In dwarf top-minnows (i.e. poeciliids) the embryos are retained within ovarian follicles which acquire an elaborate capillary bed apposed against the highly vascular embryonic yolk sac. In *Jenynsia* the ovarian lining forms highly vascular folds which contact the embryo's gills, whilst in the goodeid fishes absorptive processes grow out from the embryonic hindgut.

ENDOCRINE ORGANS

(a) Hypophysis

The teleost hypophysis is varied. In the eel a central neurohypophysis interdigitates with a surrounding adenohypophysis. This last includes an anterior pars distalis, with 'rostral' and 'proximal' components, and a posterior pars intermedia. The rostral region contains prolactin cells arranged in follicles, TSH cells as cords or masses between them, and ACTH cells at the posterior border with the neurohypophysis. The proximal region contains somatotrophs and gonadotrophs. The pars intermedia has two cell types.

(b) The thyroid gland

The thyroids of teleosts vary considerably in their rate of iodine uptake and in the proportions of thyroxin to triiodothyronine which they produce and release into the blood. Goldfish have 'slow' thyroids and only accumulate a small fraction of administered radio-iodine, whilst in salmonids there is a 5 to 10-fold increase in rate from young to 1–2 year olds. Speaking generally, thyroxin has no consistent effects on oxygen consumption but certainly affects carbohydrate metabolism, guanine metabolism and hence silvering, water and electrolyte movements, scale and bone formation and central nervous functions such as olfactory bulb and cerebral hemisphere activity. It thereby influences feeding and migratory behaviour.

(c) Ultimobranchial glands

These calcitonin-producing organs are bilaterally situated in the transverse septum between the abdominal cavity and sinus venosus. There are no follicles, the glandular tissue being diffuse.

(d) Endocrine pancreas

Three types of pancreas occur, a compact pancreas, a diffuse lobular structure, and a disseminated pancreas scattered in small patches throughout the body cavity as in *Amia* and *Acipenser*. In teleosts the Brockman bodies comprise a giant islet often separated from a rim of exocrine tissue by connective tissue. Insulin and glucagon activity have both been demonstrated.

CARDIOVASCULAR SYSTEM

i The actinopterygian system differs in a number of ways from that of selachians. A muscular conus arteriosus persists in the heart of Chondrostei and Holostei. Those of *Polypterus* and *Lepisosteus* are long and can have 7–8 transverse rows of values. In the teleosts the conus is reduced and replaced in front by a non-contractile bulbus arteriosus. It is typically represented by a narrow muscular zone with a single row of valves.

ii The persistent hyoidean hemibranch of *Lepisosteus* has an afferent vessel from the ventral aorta. This is absent from *Acipenser* and other forms where there is a pseudobranch receiving only oxygenated blood. Transverse sections of teleost gill bars reveal only two branchial vessels, an outer afferent and an inner efferent, connected by a double series of loops passing into anterior and posterior gill lamellae. The greater part of the embryonic arch forms the efferent vessel.

iii The renal portal system is variable. In *Anguilla* the caudal vein remains undivided in the posterior kidney region, then divides to send separate branches to the anterior mesonephric region. In the pike it divides to send separate branches to each kidney, and in the cod there is a large right and smaller left vessel with the right-hand one going directly to the posterior cardinal vein of its side. This right branch atrophies in salmon.

iv Various interesting ecological adaptations occur. Fishes living off Labrador exist at $-1.75°C$ and their body fluids freeze at $-1.0°C$. They are able to survive because they are out of reach of the floating ice crystals and their blood contains organic antifreeze compounds. The antarctic ice-fishes, Chaenichthyidae, lack erythrocytes but the surrounding water is high in oxygen and this is carried in physical solution in the blood.

SOUND AND LIGHT GENERATION

(a) Sound

i Water conducts sound better than air does. Less energy is lost from frictional effects and there is less turbulence so that the sound paths deviate less. Sound generation is particularly common amongst fish of both turbid tropical rivers and also the abyssal depths where vision is ineffective. Nevertheless the use of hydrophones establishes that sounds are produced by many other species during swimming, eating or expelling air from a physostomous swimbladder. An idling school of anchovy is noiseless but sounds can be detected when they are streaming or veering, and amberjack make a thumping noise as they veer. These noises are generated by the locomotion and responded to by predators, prey, or other members of the school.

ii *Notropis* species emit high-pitched sounds when gas bubbles are released from the swimbladder, and single sounds of 85 Hz to 11 kHz during the breeding season. Numerous catfishes produce sounds from the swim-

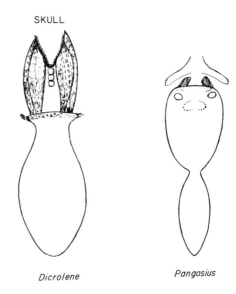

SKULL

Dicrolene   *Pangasius*

Fig. 4.14 Swimbladder with drumming muscles of the deep sea *Dicrolene* and the catfish, *Pangasius*. (After Marshall.)

bladder (Fig. 4.14). In *Auchenipterus nodosus* the two transverse processes of the fourth vertebra arc ventroposteriorly and end as an ossified plate in the swimbladder wall. The elasticity of the processes, coupled with the action of muscles originating on the skull, generates sound; the swimbladder acts as a resonating chamber. Some coastal teleosts have comparable drumming muscles. Toadfishes produce particularly loud sounds with a fundamental frequency of 140 Hz and harmonics at 140 Hz intervals up to 2 kHz. Seahorses produce snapping or clicking noises by tossing their heads when the rear edge of the skull slips under a bony crest, the coronet. *Bathygobius* generates sounds during courtship by shutting the mouth sharply and ejecting water from the operculum.

(b) Light

Below c. 750 m depth there is little or no sunlight. Many middle bathypelagic ocean fish possess photophores emitting between 410–600 mμ. These include hatchet fishes (Stomiatoidea), lantern fishes (Myctophidae) and anglers (Ceratioidea). The organs of deep-sea cod (Macrouridae and Moridae) contain luminous bacteria. Searsidae, that are related to the herrings, eject photo-generating cells into the water from a post-opercular

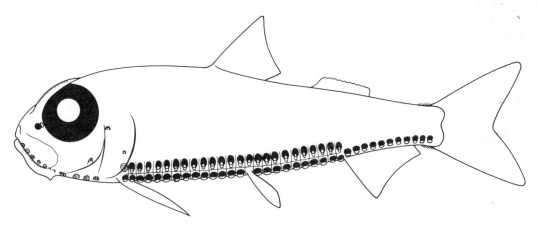

Ichthyococcus ovatus

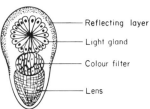

— Reflecting layer

— Light gland

— Colour filter

— Lens

Fig. 4.15 Light organs of *Ichthyococcus ovatus*, a stomiatoid fish. Note the organ in front of the eye which shines on the retina and the double row of lights along the ventrolateral region. One of these is shown below. (After Marshall.)

sac. Nearly all the stomiatoids have a species-specific arrangement of photophores comprising a pigmented sheath, silvery reflector and lens (Fig. 4.15), which are innervated. They may serve to coordinate shoals. *Diaphus* and related genera also have a large photophore on the head which illuminates the surroundings, whilst some genera have a pair by the eyes which may serve to adapt the eyes so that they are not dazzled by their own luminescence. Predatory stomiatoids such as dragon-fishes, viper-fishes, star-eaters, etc., also have a large organ in the eye region which can be retracted and hidden behind a pigment sheath.

SENSE ORGANS

(a) Olfaction

The olfactory sacs are housed in ossified capsules and the sensory epithelium is pleated, thereby increasing the surface area. There are usually two nostrils to each sac, water entering by the anterior and leaving by the posterior. The olfactory bulbs are large in bottom-living species and olfactory arousal, particularly by specific molecules, is of great importance to anadromes—and probably catadromes. Salmon exhibit specific positive olfactory responses to water from the homestream.

(b) The eyes and vision

Teleost eyes have both rod and cone cells, the latter being responsible for colour vision. The lens is spherical (Fig. 4.16) and bulges through the pupil. Light rays are not refracted and focused until they reach it and it has a very high refractive index of 1.65. Many species have a mobile iris. Double cones are common and facilitate vision in dim light. However, mormyrids, gymnotids and catfishes that are inhabitants of dimly lit water, have small eyes and rely on their electroceptors for

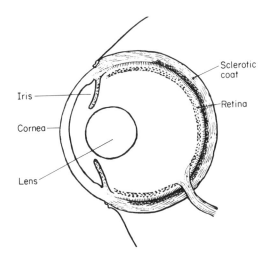

Fig. 4.16 A diagram of a salmon's eye.

perception. Reduction is extreme in the blind cave fishes such as *Anoptichthys jordani*; the congo barb, *Caecobarbus geertsi*, and the cave cyprinid of Iraq, *Typhlogarra widdowsoni*. These can all still avoid light.

Marine teleosts possess rhodopsin. Freshwater teleosts have a different pigment porphyropsin. Anadromes such as the salmon have both, and catadromes such as the eel also have both but rhodopsin predominates.

(c) Lateral line system

Lateral line canals are absent from abyssal species such as the gulper eels and ceratioid anglers in which the sense organs are on skin papillae that wave in the water. The Holostei have canals comparable with those of selachians. There is much greater variation and complexity amongst teleosts. The canals themselves are huge in Macrouridae, organs are widespread over the face in *Cobitis*, and various accessory lines occur in many families.

(d) The ear and hearing

i The ear region is essentially similar to those of selachians. In clupeomorphs diverticula from the swimbladder penetrate the skull and facilitate hearing, whilst in Ostariophysi the Weberian ossicles (q.v.) permit a similar facilitation. Many genera respond to sounds (cf. page 59), the minnow up to 4 kHz and the catfish up to 13 kHz. Pressure fluctuations of water-borne sounds make the swimbladder pulsate. If their frequency is close to the natural frequency of the gas-filled swimbladder, the amplitude of these pulsations is then enhanced. At shallow depths the natural frequency for a 20 g fish is c. 530 Hz; for a 2 kg fish 110 Hz.

ii The Weberian ossicles have slender attachments to the vertebrae and pivot freely. The posterior ones are attached to the tunica externa at the edge of a slit in the anterior chamber of the swimbladder. As the bladder expands the slit widens and the ossicles pivot forwards. An elastic ligament returns both them and the slit to their original positions as the swimbladder contracts. Their oscillations are transmitted to a fluid-filled chamber and thence to the otolith organs. In many species a gap in the somatic musculature on either side of the swimbladder reduces the damping effect of the body-wall and increases the sensitivity to sound waves close to the natural frequency of the bladder.

(e) Taste buds

Taste buds are generally restricted to the mouth and pharynx but in species with barbels, such as bichirs, sturgeons, catfishes and red mullet, etc., these also bear them. Some species have them elsewhere on the body.

(f) Touch receptors

The epidermis and dermis contain myriads of tangoreceptors. Fishes with adhesive, sucker-like pelvic fins, such as gobies, lumpsuckers, cling fishes, are no doubt guided by touch. Males of *Oryzias*, an egg-laying toothed-carp, find the females by sight and then embrace and beat them with their dorsal and anal fins. These tactile stimuli elicit oviposition. Males which have had these fins removed fail to elicit it, and females immersed in a dilute solution of anaesthetic do not extrude eggs. The tactile stimuli induce the release of hypophysial oviposition hormones.

(g) Electroceptors

The electric organs of gymnotids and mormyrids are used in conjunction with two types of electroceptors. One type resembles the ampullae of Lorenzini. The other comprises more numerous tuberous receptors. They are covered by loosely packed epithelial cells, respond to the initial and final phases of long DC stimuli and to AC stimuli up to much higher frequencies than the ampullae of Lorenzini. Electrical conductors and insulators in the environment distort the electrical field and greatly affect the number of action potentials generated.

THE BRAIN

(a) The medulla

The structure of the medulla is intimately associated with the relative predominance of different sensory modalities. Acoustic responses are usually widespread in the dorsal regions but, for example, in cypriniform and siluriform genera the taste receptors on the cephalic and trunk epidermis are associated with large facial, glossopharyngeal and vagal aggregations.

Two Mauthner cells occur in the reticular formation and always mediate swift myotomal responses to anterior stimuli, although their precise functions vary from species to species. Neurons controlling the respiratory cycle of mouth and operculum occur in a strip on either side of the midline. In the weakly electric gymnotids a small number of medullary pacemaker cells initiate the discharge.

(b) The cerebellum

Actinopterygian cerebella are very variable but typically comprise a central corpus cerebelli, two lateral auricles and an anterior prolongation, the valvule. Those of mormyrids are huge and hypertrophy of the valvule gives a series of foliate structures that cap the rest of the brain (Fig. 4.17). Their long-suspected electrosensory involvement has recently been confirmed.

(c) The mesencephalon

The optic tectum is very complex in species for which the visual modality is important. It has a precise retinotopic organization (Fig. 4.18), on and off responses occur, and the units have well-defined visual fields. The ascending

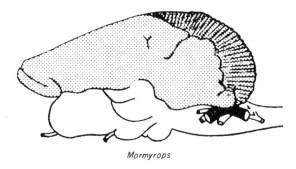

Fig. 4.17 The greatly enlarged cerebellum of the electroceptive fish *Mormyrops*.

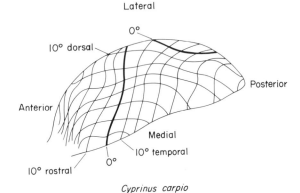

Fig. 4.18 The optic tectum of *Cyprinus carpio* showing the representation of the meridians of the visual field. (After Schwassman & Kruger.)

auditory pathway relays at the torus semicircularis and the dorsal part of this is involved, in electric fishes, with electrical stimuli of social significance.

(d) The diencephalon

The diencephalon is a very complex brain region with numerous nuclei whose prominence varies from species to species. The roof bears the pineal organ and the thalamus is particularly complex. Numerous feeding and aggressive responses have been elicited by stimulating the preoptic region which is also a site of hypophyseal controls.

(e) The cerebral hemispheres

The actinopterygian cerebral hemispheres differ from those of all other vertebrates. In other classes they develop as lateral evaginations of a median structure. In teleosts the hemispheres result from eversion (Fig. 4.19) of the lateral walls of the embryonic neural tube and the complex nuclear aggregations are, therefore, difficult to homologize with those elsewhere. Numerous fibre tracts communicate with other brain regions. Experimental investigations demonstrate hemisphere involvement in arousal and learning, but studies involving hemisphere extirpation are difficult to evaluate as extensive or even total regeneration occurs. Various regions are involved in the initiation and integration of complex behaviour, such as courtship and parental behaviour in sticklebacks, and localized interference in the caudal hemisphere regions exaggerates or eliminates the various components of such behaviour.

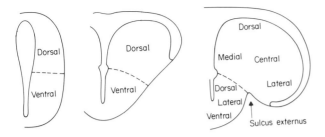

Fig. 4.19 Diagrams showing the developmental phases of eversion in a teleost forebrain.

FOSSIL HISTORY

Actinopterygian fishes first evolved at some undetermined time in the mid-Paleozoic. Six well-documented paleoniscoid genera are known from the Middle and Upper Devonian, but only scales of questionable affinities occur in older deposits. The fossils come from both marine and freshwater deposits which has led to much controversy about whether they were all freshwater with some washed out into marine environments; whether some were marine and others were freshwater, or whether some were euryhaline. The data do not permit definitive conclusions. As there was extensive diversification during the Carboniferous and Permian, the precise interrelationships of the genera themselves, and their relationships to later forms, are difficult to assess. However, in broad terms the Permian and Triassic waters were dominated by a wide variety of chondrostean fishes. During the Jurassic these were replaced by holosteans and many late Permian and early Mesozoic forms also have features suggestive of holosteans. These produced extensive radiations of so-called sub-holosteans as well as the extensive holostean radiations which dominated Jurassic waters. In the Cretaceous they were gradually replaced by the teleost lineages.

DIVERSITY OF TELEOSTS

i Teleosts are astoundingly diverse and adapted for life in all the available aquatic niches. Greenwood and his collaborators proposed three principal taxonomic groupings and were clear that various separate evolutionary routes lead independently towards an acanthopterygian grade of organization. Their first grouping included the elopomorph and clupeomorph fishes. They assumed that in this grouping evolution from a primitive elopiform ancestry had produced eel-like and herring-like forms (Fig. 4.20). Characteristic and primi-

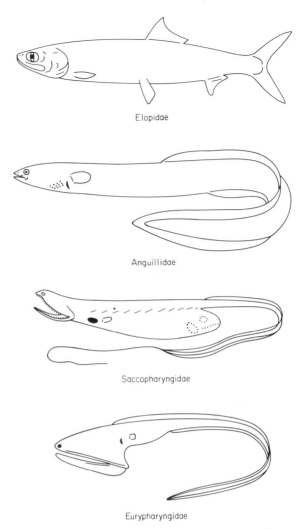

Elopidae

Anguillidae

Saccopharyngidae

Eurypharyngidae

Fig. 4.20 Various elopomorph families showing their morphological diversity.

tive characters include a short, broad maxilla with a large movable supramaxilla; maxillary teeth that are seldom excluded from the gape; parasphenoid and pterygoid teeth, and a functional pneumatic duct. Eels have a leptocephalous larva.

ii The second grouping comprises the osteoglossomorphs (Fig. 4.21). These all retain primitive traits in their jaw suspension and pectoral girdle. The maxillae, toothless in mormyrids, contribute to the gape, and there are parasphenoid, glossohyal and pterygoid teeth. There is a functional pneumatic duct and the caudal fin tends to be reduced and confluent with the anal and dorsal fins.

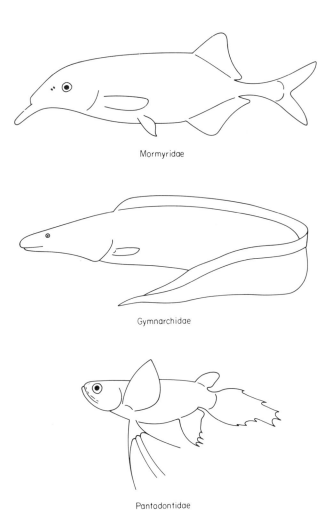

Fig. 4.21 Morphological diversity of osteoglossomorphs.

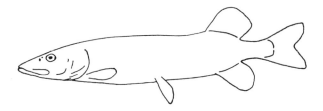

Fig. 4.22 The pike, a member of the immensely diverse third grouping of teleost families.

They are predominantly tropical and freshwater fishes of insectivorous or piscivorous habits. There are two orders, the Osteoglossiformes, including the Arapaima,

and the Mormyriformes including the mormyrids and gymnarchids.

iii The third grouping includes five superorders. They are usually of a distinctly teleostean appearance (Fig. 4.22), and characteristic trends include the loss of the pneumatic duct, a lowering of the centre of gravity and the approximation of the centre of buoyancy to the centre of mass. The large and frequently mobile premaxilla partially or completely excludes the maxilla from the gape and the maxillae, parasphenoids and pterygoids lose their teeth. Pectoral fins assume a position high on the side of the body, the pelvic fins tend to migrate forwards, and the pelvic girdle links up with the pectoral. There are progressive reductions in the number of fin bones and the hypural bones are reduced to a single unit. Several primitive lines develop an adipose fin and various lines develop fin spines and ctenoid scales. The Ostariophysi develop Weberian ossicles.

The first of the five superorders, the Protacanthopterygii, are predominantly slender, predatory fishes. Photophores occur in oceanic forms. The upper jaw is slightly protrusible and an adipose fin is usually present. The second superorder, the Ostariophysi, is predominantly freshwater. It includes predatory, herbivorous, toothless detritus-eating and microphagous forms. All have Weberian ossicles, the upper jaw is frequently protrusible and fin spines are also often present. An adipose fin is often present, scales can be present or absent, are usually cycloid, but are ctenoid in a few genera.

The Paracanthopterygii are mostly marine, soft-bodied, carnivorous, deep-water genera. There is a virtual loss of photophores, and cephalic mucous canals are prominent. Modified ribs link the skull to the cleithra in many species. The pelvic fins are anterior and the swimbladder is frequently subdivided and connected by diverticula to parapophyses of precaudal vertebrae. There are also sometimes ossicles linking it to the ear. The Atherinomorpha are generally small, surface-feeding fishes of fresh and brackish water that exhibit pronounced sexual dimorphism. Many males have bony external genitalia derived from the anal, pelvic or pectoral fins and there are numerous viviparous species.

Finally, there is the superorder Acanthopterygii. These are principally marine and benthic genera of very varied appearance. Photophores are uncommon and the upper jaw is protractile. The bones of the head often have strong spines, fin spines are also usually present and the opercular apparatus is frequently similarly armed. If present, the pelvic fins have a pectoral or jugular location. Hypural bones always emanate from a single centrum. Connections between the swimbladder and ear are rare and never involve ossicles.

FURTHER READING

ALEXANDER R.McN. (1967) *Functional design in fishes.* Hutchinson University Library, London.

GOSLINE W.A. (1971) *Functional morphology and classification of teleost fishes.* University Press of Hawaii, Honolulu.

GREENWOOD P.H., MILES R.S. & PATTERSON C. (1973) *Interrelationships of fishes.* Linnaean Society and Academic Press, London.

HARDEN-JONES F.R. (1968) *Fish migration.* Edward Arnold, London.

HOAR W.S. & RANDALL D.J. (eds) (1969) *Fish Physiology*, 7 vols, Academic Press, London.

KLEEREKOPER H. (1969) *Olfaction in fishes.* Indiana University Press, Bloomington and London.

MARSHALL N.B. (1965) *The life of fishes.* Weidenfeld and Nicholson, London.

MARSHALL N.B. (1971) *Explorations in the life of fishes.* Harvard University Press, Cambridge, Mass.

NORMAN J.R. & GREENWOOD P.H. (1963) *A history of fishes.* Ernest Benn Ltd., London.

5 · Class Osteichthyes 2. Crossopterygii and Dipnoi

INTRODUCTION

The lungfishes, coelacanths and fossil rhipidistians remain to be considered. They are only represented in the modern world fauna by three genera of lungfish and one species of coelacanth. However, all three groups are widely represented by fossils and, like the paleoniscoid fishes, are known from the Devonian. Their interrelationships, and their relationships with other fish groups, remain controversial. The lungfishes and coelacanths are widely viewed as divergent lineages whose origins lie amongst early Devonian rhipidistians but other suggestions see them as quite independent taxa. Some authors who adhere to such views consider that *Polypterus* and *Calamoichthys* are unrelated to the actinopterygian assemblage and constitute a further independent osteichthyan lineage. Not surprisingly such contrasting suggestions are associated with differing conclusions about the relative proximity of the coelacanths and Dipnoi to the ancestry of tetrapods. This is itself possibly polyphyletic and all the groups are best viewed as exhibiting both homologous and parallel adaptations to late Devonian environments.

Crossopterygii

Rhipidistia

The rhipidistians comprise a relatively homogeneous fossil assemblage, members of which were amongst the first Osteichthyes to diversify in the stratigraphical series. *Porolepis* of the Lower Devonian is indeed, the earliest known osteichthyan. They are thought to have been largely freshwater forms. *Osteolepis* and *Eu-*

Eusthenopteron

Fig. 5.1 The fossil genus *Eusthenopteron*. Note the symmetrical tail with its epichordal and hypochordal lobes.

sthenopteron (Fig. 5.1) exemplify the general morphological organization. There were two dorsal fins and one anal, and the body was covered with regularly organized cosmoid scales in which bony isopedin is overlain by dentine-like cosmine and there is a thin layer of enamel-like ganoin externally.

SKELETAL SYSTEM

(a) The skull

i The cranium was well ossified and covered by a complete roof of dermal bones bearing canals for the lateral line system. It is possible that ossification was intimately associated with this system as bone growth occurs around the lateral line in teleosts, excision of the line negating ossification and implantation initiating it. In front of the orbits the skulls of early genera included a mosaic of small bones that are perhaps relics of a more extensive mosaic encompassing the entire skull. Further back the bones of known forms were, however, larger, and their pattern is reminiscent of that in early tetrapods.

ii In *Eusthenopteron* (Fig. 5.2) the front of the roof included nasals, rostrals and a postrostral. The frontals did not meet in the midline as they were separated by this postrostral. Both the parietals, with the pineal foramen between them, and the postparietals of each side were apposed in the midline. A posterior transverse assemblage included an unpaired median extrascapular

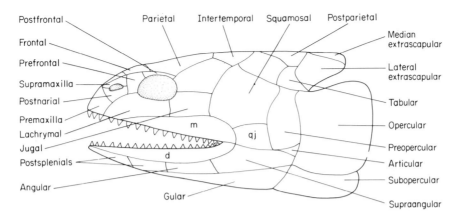

Fig. 5.2 Lateral view of the skull of *Eusthenopteron*—m, maxilla; d, dentary; qj, quadratojugal.

and paired lateral extrascapulars. The orbits were generally surrounded by lachrymal, prefrontal, post-frontal and jugal elements, although in more recent genera this last is situated more ventrally and between the maxilla and quadratojugal. Dorsal supratemporals and tabulars, and lateral squamosals and preoperculars occupied the otic regions. The upper jaw comprised palatines, metapterygoids and quadrates with premaxillae in front and quadratojugals behind. Meckel's cartilage was ossified posteriorly giving the articular, and covered by a long dentary of dermal origin over much of its external surface. Below and behind this were splenial, postsplenial, angular and supra-angular elements. Most of the internal surface was covered by prearticulars with small coronoid bones on the upper edge.

iii The outstanding functional attribute was an intra-cranial joint. An anterior cranial assemblage in front of the notochord articulated with a more posterior assemblage penetrated by the notochord. The palatoquadrate was hinged to the first-named, the hyomandibular to the last-named with a spiracle between them. As the mouth opened the hyomandibular swung laterally, pushing the jaws forward and displacing the anterior cranial component upwards, this gave a wider gape. Teeth were borne on the premaxillae, maxillae, palatines, vomers and ectopterygoids, and, in the lower jaw, on dentary and coronoids. They were labyrinthine in form (Fig. 5.3) as were those of early amphibians.

iv Some had two external nostrils, some only one. All had internal nostrils—choanae.

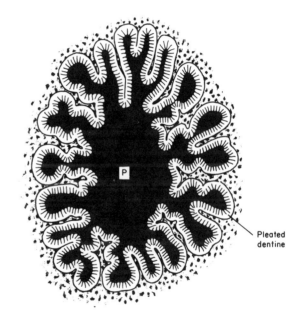

Fig. 5.3 A section through a tooth of *Eusthenopteron* showing the pulp cavity (P) and pleated dentine. (After Schulze.)

(b) The axial skeleton

The vertebral column tended to be poorly ossified. As in early amphibians (see page 75) the notochord was often surrounded, except on its dorsal side, by hypocentra, and the small pleurocentra lay on its dorsal side between the hypocentra. Neural arches were attached to the dorsal edges of the hypocentra, surrounded the nerve cord, and bore dorsal neural spines to which the epaxial muscles were attached. Ribs articulated almost entirely with the hypocentra, having only small areas of apposition on the neural arches. Early genera had heterocercal tails with a small, dorsal, epichordal lobe. This then increased in size during the Devonian to give the characteristic, symmetrical, three-lobed, gephyrocercal tail. Such symmetry implies that there was no longer a need for the heterocercal condition in order to maintain a constant depth and presumably reflects the presence of either an airbladder or wax esters comparable to those in coelacanths (cf. below).

(c) Girdles and appendicular skeleton

i The pectoral girdle was comparable with that of paleoniscoid genera. It formed a complete circle around the body behind the operculum and provided an origin for the paired fin and body musculature, as well as strengthening the body behind the weak point represented by the gill region. Scapulocoracoid elements were accompanied by paired clavicles that met in the midline and, above these, cleithra and supracleithra, the last-named articulating with the post-temporals of the skull. The pelvic girdle, consisted of a single pair of bones.

ii The structure of the paired fins is shown in Fig. 5.4. There was a bony axis of pterygiophores with radial elements on each side. This is the unfortunately des-

Fig. 5.4 The so-called archipterygial fin-type.

ignated archipterygium. Long and leaf-like in *Holoptychius* they were short and ovoid in *Osteolepis*.

Actinistia

General Characteristics

The Coelacanthini first appear in the Upper Devonian and resemble the contemporaneous Rhipidistia. Long thought to have been extinct since the Cretaceous, the discovery of the living *Latimeria chalumnae* (Fig. 5.5) showed that they continued to exist and hence aroused immense interest. They have a hinged skull, cosmoid scales, a gephyrocercal tail, and lobed fins. The most obvious differences from most Rhipidistia are the shorter, stockier body, with a relatively short mouth and skull, and the reduction of the maxilla and dentary. Other differentiating characters are the absence of a

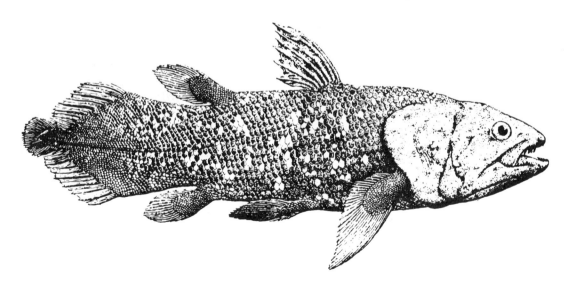

Fig. 5.5 The modern coelacanth genus *Latimeria*. (From Millot.)

pineal foramen, the presence of a large organ of questionable function in the snout above the mouth, and the absence of choanae or internal nares. An airbladder is ossified and preserved in the Cretaceous genera *Macropoma* and *Undina*. *Latimeria* inhabits abyssal regions and, as a result, the absence of a lung is hardly surprising, but the ancestral gut diverticulum is represented by an organ, containing wax esters, whose internal cavity is a narrow oesophageal diverticulum. Such esters are also widespread in the musculature and contribute to buoyancy.

The upper jaw is autostylic and attached to the anterior part of the skull, but a basipterygoid articulation between the auditory region and the metapterygoid is absent, perhaps permitting greater freedom of movement of the front part of the skull on the intracranial hinge. Manipulation of the head of dead specimens shows that lateral displacement of the palatoquadrates fails to flex the intracranial joint as it is assumed to have done in Rhipidistia. Opening the mouth does, and the jaw is pushed forward.

VISCERA

In the body cavity the majority of organs resemble those described for actinopterygians. A large stomach is separated from the intestine by a sphincter and the intestine has a spiral valve of twenty turns. Liver, pancreas and spleen are well-developed. The kidneys are, however, aberrant. Fused to form a Y-shaped structure they lie on the ventral wall of the body cavity behind the anus. The heart retains a broadly longitudinal organization with the sinus venosus and auricle behind the ventricle and not above it. As it combines primitive features with many that are common to Crossopterygii generally, and yet others that are peculiar to itself, *Latimeria* confirms the general conclusions previously drawn by paleontologists on the basis of fossil material.

Dipnoi

General Introduction

The lungfishes represent a separate subclass of the Osteichthyes. From their origin in the Middle Devonian, lungfishes again persisted as common forms until the Cretaceous. The modern genera *Protopterus*, *Lepidosiren* and *Neoceratodus* (Fig. 5.6) inhabit tropical rivers and swamps in Africa, South America and Australia respectively.

THE SKULL

i Despite the important diagnostic features, such as lobed fins, that suggest affinities with the crossopterygians, even the earliest fossil genera exhibit striking differences in both skull and dentition. The oldest have a skull roof which is composed of a mosaic of bones that are again particularly numerous anteriorly. There is then a marked trend towards reducing the number of roof bones. Cosmine and enamel layers may be present in *Dipterus* but these outer layers are absent in more recent genera where the anterior cephalic region lacks dermal bones. The lateral walls become reduced and, in extant genera, only a few dorsal plates remain.

ii Cartilaginous elements of the jaws and gill arches exhibit little ossification even in early forms. In recent genera the primary upper jaws have fused to the braincase and there is a concomitant reduction of the hyomandibular which is involved in the jaw suspension of other fish.

THE DENTITION

The teeth of lungfish are characteristic. Marginal teeth only occur in a few Devonian and Carboniferous

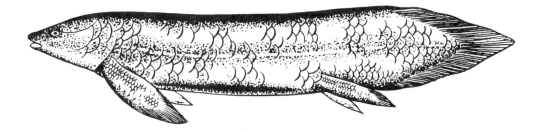

Fig. 5.6 The Australian lungfish *Neoceratodus*.

genera. Some Devonian genera have thickened bony rims at the front of both the upper and lower jaws which probably functioned as cropping devices. Evolutionary developments then accentuate the formation of tooth-plates close to the midline. The prevomers usually bear small toothplates but the principal structures are carried by the pterygoids above, and the prearticulars below. In the fossil genus *Dipterus* these bones bear radial rows of prominent teeth. In more recent genera such teeth are fused to form connected ridges borne on thickened, fan-shaped plates (Fig. 5.7). These are adapted for shearing and crushing small molluscs and other invertebrates.

RESPIRATORY SYSTEM

i The young of *Protopterus* and *Lepidosiren* use gills as organs of gaseous exchange. *Neoceratodus*, with a single lung, also relies on gills for much of its gaseous exchange when, as an adult, it is in well-aerated water. Breaths through the mouth are then usually about one hour apart. In low oxygen tensions water is pumped faster over the gills and the breaths are more frequent. Adult *Protopterus* and *Lepidosiren*, with two lungs, have reduced gills. Indeed *Protopterus* lacks gill filaments on two gill bars. Even in highly aerated water they go to the surface every 3–10 minutes. The lung walls are highly vascularized and the air is only separated from the blood by the capillary walls and surface epithelium representing a thickness of about 0.5 μm.

ii X-ray cinematography shows that the fish closes its mouth whilst still submerged and drives water out from the buccal and opercular cavities through the opercular openings. The opercula are then closed by a transverse muscle in the throat and held closed until air in the lungs has been changed. The mouth is pushed through the water surface and opened. At this stage the lungs empty. The mouth is closed, the floor of the buccal cavity raised, and, as the air cannot escape via the opercula, it enters the lungs.

VASCULAR SYSTEM

Oxygenated blood from the lungs, and deoxygenated blood from the rest of the body, are kept fairly separate in the heart. Blood is carried to the lungs by a pulmonary artery that arises from the last efferent branchial vessel. It returns directly to the heart via a pulmonary vein which enters the left side of the sinus venosus at an opening which is distinct from that used by other venous blood. The two types of blood are then prevented from mixing by ridges on the internal walls. These form septa that broadly separate the chambers into right and left parts, and continue into the ventral aorta as a horizontal partition. The oxygenated blood is, therefore, channelled to the anterior branchial arches and thence to the dorsal aorta and general body circulation. The fact that blood does not proceed directly from the lungs to the body in general presumably reflects the high arterial pressures which would be necessary to drive it through several capillary networks. These would give a far greater pressure difference between the blood and air in the lungs and necessitate stronger and, therefore, thicker pulmonary capillary walls and epithelia to prevent them bursting.

AESTIVATION

The swamps inhabited by *Protopterus* dry out in the dry season. At this time of year *Protopterus* takes mouthfuls

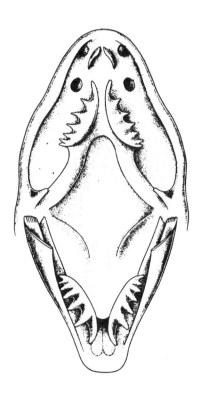

Fig. 5.7 The mouth of *Neoceratodus* showing the toothplates.

of the wet mud, ejects it through its gills, thereby burrowing. As the mud dries out the mucous coating of the skin forms a thin coccoon which prevents desiccation. This coccoon formation is not shown by *Lepidosiren*, and *Neoceratodus* does not aestivate. The ability to aestivate is, however, clearly a long-established character, as fossil dipnoan burrows are well known from the Devonian and Carboniferous. *Protopterus* normally excretes ammonia. Although typical of aquatic forms, this would be lethal in the absence of a copious diluent, and during aestivation urea is accumulated.

FURTHER READING

ALEXANDER R.Mc. (1975) *The chordates*. Cambridge University Press.

FORSTER G.R. (1974) The ecology of *Latimeria chalumnae*. *Proc. R. Soc., Lond.* **B. 186**; 291–6.

MILLOT J. & ANTHONY J. (1958) Crossopterygians actuelles. In Grassé P.P. (ed.) *Traité de Zoologie*, vol. 13., fascicule 3; 2553–97.

NORMAN J.R. & GREENWOOD P.H. (1963) *A history of fishes*. Ernest Benn, London.

THOMPSON K.S. (1969) The biology of the lobe-finned fishes. *Biol. Rev.*, **44**, 91–154.

6 · Class Amphibia

INTRODUCTION

The earliest known amphibians are the Ichthyostegalia of late Devonian and early Carboniferous deposits. They exhibit many features that are intermediate between those of their supposed rhipidistian ancestors and those of more recent genera. During the Carboniferous and Permian periods many diverse forms then evolved. Modern amphibians are grouped in three orders. The Urodela, salamanders and newts, retain a primitive elongated form with short legs, long trunk and well-developed tail. The Apoda or Gymnophiona comprise the legless, burrowing caecilians that are confined to the tropics, and the Anura includes all frogs and toads. These three orders are usually placed within a subclass Lissamphibia. All share such features as reduced ribs and, when legs are present, unossified carpus and tarsus, together with an amphibian papilla in the ear and pedicillate teeth. None have more than four toes in the front feet. Some authorities separate the anuran stock from the urodelan one, envisaging the similarities as the results of convergent evolution.

INTEGUMENT

i The skin of all tetrapods plays important roles in both protection and water conservation. In the terrestrial adults of modern amphibian genera this last-named role conflicts with the emphasis on cutaneous respiration and most species inhabit humid environments. The earliest fossil genera had a covering of scales. Rudimentary dermal scales still occur buried in the skin of some caecilians, and a few toads have bony plates embedded in their backs. In deserticolous genera, such as *Chiroleptes*, the top of the head can be cornified, reducing water loss when this is the only part of the body exposed to the environment at the top of a hole or burrow. Furthermore, *Xenopus laevis* has cornified epidermal structures on the first three digits of the hind limb. Nevertheless most modern amphibians lack scales and have a soft skin.

ii Amphibian skin includes the usual epidermis and underlying dermis (Fig. 6.1). The former consists of several cell layers. Horny, dead cells of the outermost stratum corneum are shed periodically either in bits, as in *Rana pipiens*, the leopard frog, or as a whole, as in *Bufo*. Not present in fish, and lacking from a few perennibranchiate urodeles, the outermost cornified cells protect the underlying living tissue and must make at least a small contribution to water conservation. The dermis is usually relatively thin. An outer, loosely organized stratum spongiosum overlies an inner, more compact, stratum compactum. It serves as an organ of gaseous exchange and is usually well vascularized. Pigment is contained in special cells, chromatophores, that are located in the uppermost layer of the dermis, or between the dermis and epidermis.

iii Unicellular glands, common in fishes, are largely restricted to patches on the head of the embryo and secrete enzymes that assist hatching by dissolving the gelatinous envelopes of the eggs. Some larval salamanders have other glands whose function remains unknown. Multicellular mucous and poison glands are numerous and widespread. The mucous secretion helps to keep the skin moist, thereby providing a medium for gaseous exchange. Poison glands vary in their occurrence but are well-developed in many terrestrial anurans. The warts of toads, and the parotoid glands on

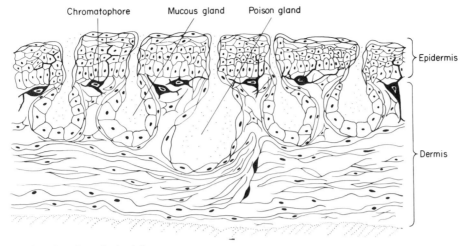

Fig. 6.1 Diagram showing the principal features of amphibian skin.

the back of the neck, have large numbers of them. The poisonous products are frequently lethal to other species and are generally of great selective value in rendering the animals distasteful to potential predators. This is frequently associated with the presence of warning colouration. Many, such as those of *Hyla vasta*, a large tree-frog from Haiti, irritate human skin. Perhaps the best known is bufotenin, 5-hydroxydimethyl-tryptamine, which causes vomiting and convulsions in humans.

Modified integumentary glands also serve a number of other functions. Glandular discs at the tips of the digits of tree-frogs such as *Hyla* and *Leptopelis* assist adhesion and the glandular thumb pads developed by many male anurans during the breeding season help them to clasp the females. Male urodeles have glands on both the face and tail whose secretions stimulate the females when wafted towards them by male tail movements.

iv The skin of fishes bears various chemo- and mechano-sensory organs. In general these are also present in the moist skin of Amphibia. Lateral line systems are present in most larvae and the adults of many permanently aquatic genera. They are lost at metamorphosis in other genera whose adults are terrestrial but persisted in fossil genera.

SKELETAL SYSTEM

(a) Cranial region

i In ichthyostegalians the bones of the cranial roof

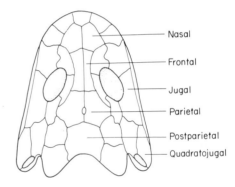

Fig. 6.2 The skull of *Ichthyostega*.

were closely similar to those of rhipidistians (Fig. 6.2). Differences include a reduction of the overall length; the replacement of the spiracular cleft by a shallow otic notch, and the absence of both intertemporal and median extrascapular bones and of an intracranial articulation. A suture behind the hypophysial foramen parallels the articulation of rhipidistians. Various features common to the skulls of both taxa are absent from those of more recent amphibians. A small rostrolateral bone bearing a component of the infra-orbital sensory canal occurs in ichthyostegids and is homologous with the rostrals of osteolepids. Small preopercular bones are present at the posterolateral angles and behind them there are small suboperculars.

ii Subsequent evolution within the labyrinthodont assemblage produced two divergent types of organization epitomized by the embolomerous and rhachitomous lineages (cf. Fig. 6.7). The skulls of the early genera within both lineages had many features in common and retained various archaic characters mixed with more advanced ones. Carboniferous embolomeres such as *Paleogyrinus* (Fig. 6.3) had large tabulars and persistent intertemporals. Some early rhachitomes (Fig. 6.4) also had intertemporals as well as septomaxillary bones, whose homologies are doubtful, in the region of the nostrils. A lachrymal duct, possibly present in ichthyostegids, drained the orbit into the nasal passage. Behind the parietals the otic region of the roof was shorter than in ichthyostegids and the otic notch was deeper. A supraoccipital ossification in the cartilage of the hind wall of the skull paralleled the actinopterygian condition.

Ventrally the skulls of both groups were similar to those of ichthyostegids and rhipidistians. Amongst rhachitomes the pterygoid and ectopterygoid bones were large and covered most of the ventral surface in front of the braincase. In the earliest representatives their inner edges were almost in contact with the median parasphenoid. There was hardly any interpterygoid vacuity. What appears to have been a mobile articulation occurred between the pterygoids and basisphenoid—the basipterygoid articulation. Where present, the lateral line canals no longer lay within the bones, as in ichthyostegids, but rather in grooves along their surfaces.

iii The Seymouriamorpha, long thought to be reptiles until specimens of *Kotlassia* were found with impressions of external gills, had a dermal roof of the embolomerous type (Fig. 6.5). However, the single condyle, formed by basioccipital and exoccipitals, was rounded like those of reptiles not flat as in other labyrinthodonts. There was no interpterygoid vacuity and the parasphenoid and basisphenoid were fused into a large plate lying below the braincase. They appear to have been right at the transition point to a reptilian type of organization.

iv Evolution within the labyrinthodont assemblage involved a number of trends. Thus in temnospondyls:
    the articulation between the pterygoids and basisphenoid became stronger and wider;
    the interpterygoid vacuities enlarged;
    the occipital condyle became double as the basioccipital was reduced;
    the skull became flattened with an epipterygoid ossification forming a strut between the prootic and pterygoid;

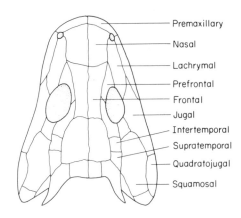

Fig. 6.3 The skull of *Paleogyrinus*, an embolomere.

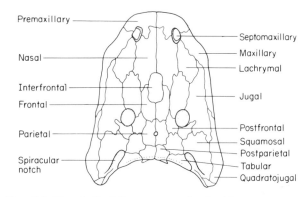

Fig. 6.4 The skull of *Eryops*, an early rhachitome.

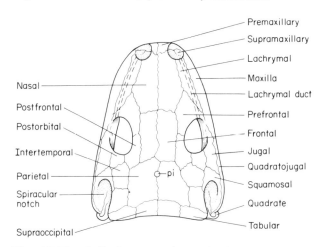

Fig. 6.5 The skull of a seymouriamorph. pi, pineal foramen.

a progressive loss of ossification within the braincase was compensated for by an enlargement of the dermal bones which grew downwards at the back of the skull.

The flattening of the skull enlarged the volume of the buccal cavity and probably improved the pulmonary gaseous exchange mechanism (see below). Interpterygoid vacuities may have allowed retraction of the eyes into the mouth region as in modern forms. In addition, the jaw articulation, which was originally far behind the occipital condyle, moved forward and upwards, and the lower jaw became prolonged backwards behind its articulation. This increased the leverage and facilitated the action of the mandibular depressor muscle that originated on the skull and was inserted on the hind part of the lower jaw.

v Modern Anura have a very different skull due to the loss of many dermal bones and the fact that the remainder have sunk below the skin of the head so that this is soft. The frontals and parietals are fused to a single median frontoparietal unit (Fig. 6.6) and a large sphenethmoid bone surrounds the front of the braincase. Various other specialized forms have occurred in the past. Many so-called microsaurs combined reduced ossification of many bones with greatly enlarged supratemporals which covered much of the hind part of the skull. Nectridia, short-legged, elongated lepospondyls from the Carboniferous and Lower Permian, bore large horns on the otic region.

(b) Axial skeleton

i The development of the vertebral column in modern amphibians differs somewhat from that in Chondrichthyes but as the rhipidistians have similar columns it is presumed that theirs developed in a similar way. The early sclerotome again gives rise to cranial and caudal half-segments. Primitively two centra are then formed in each vertebra, an anterior hypocentrum from the caudal half of the anterior sclerotome, and a posterior pleurocentrum from the cranial half of the next sclerotome. Additional contributions to the hypocentrum are de-

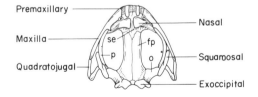

Fig. 6.6 The skull of an anuran. Abbreviations: fp, frontoparietal; o, prootic; p, pterygoid; se, sphenethmoid.

rived from tissue in the cranial half of its sclerotome, whilst the pleurocentrum has contributions from both halves of its sclerotome and from the caudal half of the more anterior sclerotome. Furthermore only these two centra, and the neural arches above the hypocentra, occur. No ventral arcualia are formed. Haemal processes, and arches if they are present, are formed by ventral extensions of the hypocentrum. The centra themselves form outside the notochordal sheath, the dorsal arcualia make no contribution to them, and they are, therefore, called autocentra not arcocentra. The neural arches bear zygapophyses which articulate with the neighbouring arches. These are derived from the half sclerotomes in front and behind that from which the neural arch itself is derived. Whether those of actinopterygians have a similar origin is unclear.

ii The ichthyostegalian column was essentially similar to that of rhipidistians although the dorsal spines were vertical and the individual neural arches articulated by way of zygapophyses, giving a greater strength and rigidity. A tail fin was borne behind and supported by definitive pterygiophores. The points at which the ribs articulated on the hypocentrum and neural arches were separated by a small space, and such a double articulation is characteristic of terrestrial vertebrates. That on the hypocentrum is known as the capitulum, that on the arch, the tuberculum. The vertebrarterial canal passes through the intervening space.

iii Divergent evolutionary trends are exhibited by the vertebrae of the rhachitomous and embolomerous lineages (Fig. 6.7). Rhachitomes have the more archaic arrangement, recalling that in ichthyostegids. There were large wedge-shaped hypocentra and smaller dorsal pleurocentra. The notochord remained large. Subsequent evolution within the rhachitomes involved a reduction in the size of the pleurocentra and these are absent from stereospondylous vertebrae. The hypocentra persist as strong cylinders that surround the notochord, which is constricted at their centres.

The embolomeres have well developed hypo- and pleurocentra in the form of solid discs. Seymouriamorph vertebrae can be derived from this condition by reduction of the hypocentra and enlargement of the pleurocentra. The pleurocentra form almost all the vertebrae, with the hypocentra remaining as small wedges receiving the rib capitulum. Both the embolomerous and rhachitomous trends restrict the notochord and replace it by the centra as the main skeletal elements. This gave the additional strength needed for

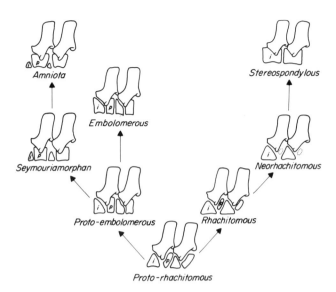

Fig. 6.7 Diagram showing the putative phylogenetic relationships of different amphibian vertebrae and the origin of the amniote condition. i, hypocentrum; p, pleurocentrum. (From Romer.)

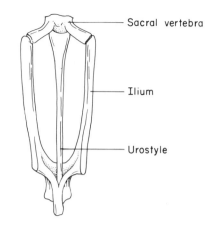

Fig. 6.8 The urostyle and pelvic girdle of a frog.

life on land which lacks the supportive effects of water. In the Nectridia and snakelike aistopods there was simply a single, complete, bony centrum around the notochord.

iv The structure of the centrum in frogs and toads is controversial. Some people consider it to be a ventral extension of the neural arches, others that it is a pleurocentrum. The vertebral column itself is also greatly foreshortened, there being only six presacral vertebrae in some frogs. The post-sacral vertebrae are fused into a single bone, the urostyle (Fig. 6.8). Movable ribs are absent but the large transverse processes of the neural arches may incorporate rib components. An overall reduction in the ossification of the skeleton occurs in genera such as *Telmatobius* which are the denizens of deep lakes.

(c) The girdles

i The pectoral girdles of ichthyostegalians lacked a connection with the skull although one was provided by a horn on the tabular in some embolomeres. Its absence perhaps reflects the relatively stronger body that resulted from loss of a gill region. There were large cleithra and, below these, small clavicles with a median inter-clavicle between them. The cleithra were also large in

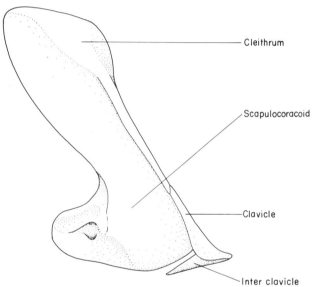

Fig. 6.9 The shoulder girdle of *Eryops*, a rhachitome. (After Romer.)

embolomeres but reduced in rhachitomes (Fig. 6.9). They were particularly small in seymouriamorphs, where the clavicle was large. Ossification of the cartilage bones to give a dorsal scapula and two ventral bones, the coracoid and procoracoid, again intimate a close relationship between these animals and reptilian lineages.

ii In modern Amphibia few remnants of the cleithral arch remain although a trace of it has been said to lie at the anterior edge of the frog's scapula. In Urodela the

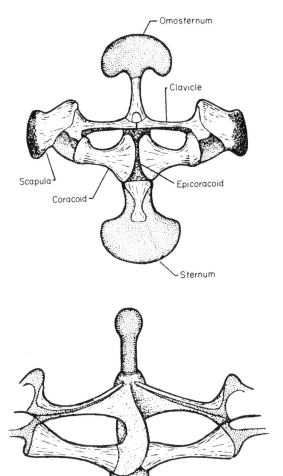

Fig. 6.10 Anuran pectoral girdles. Firmisternous above and arciferous below.

In the saltatory frogs the inner border of the coracoid is attached to its partner on the other side. This limits the mobility of the shoulder joints, and in some genera the clavicles, no longer acting as connecting rods due to the fixation of the coracoids, degenerate. This firmisternous condition is presumably an adaptation for jumping. The sternum is lost in Apoda but is a small plate in urodeles and arciferous anurans. It can have an ossified osmosternum attached to the girdle in firmisternous genera.

iii The pelvic girdle of ichthyostegids resembled that of fishes as it comprised a single bone on each side, but this was much smaller than those of typical fish. Complementarily, it also resembled that of later Amphibia as it extended dorsally to articulate with a sacral rib. In rhachitomes there are three bones, a dorsal ilium and ventral ischium and pubis (Fig. 6.11), and this condition persists in all later tetrapods. The rhachitome ilium articulated with a single sacral vertebra.

In seymouriamorphs the girdle was strong, expanded ventrally into a broad plate, had a very strong articulation with one sacral vertebra and a tendency to articulate with a second. In anurans the ilium articulates with the transverse process of a single sacral vertebra and in some frogs this ilio-sacral articulation is movable. The column is able to rotate in a sagittal plane and, in *Xenopus*, to slide backwards and forwards on the ilium.

(d) Pentadactyl limbs

i The limbs consist of chains of bones that are divisible into four principal sections (Fig. 6.12). A powerful proximal element, the humerus or femur, articulates with the pectoral or pelvic girdle by a joint which

clavicles are lost and the girdle does not ossify. Two principal types of pectoral girdle are found in Anura (Fig. 6.10). Ambulatory genera, such as toads, have arciferous girdles. Here the coracoid plate has a large fenestra dividing it into pro- and postcoracoid elements. The procoracoid is fused to the clavicle which is attached at its distal end to the scapula, and hinged at its median end to either an anterior element of the sternum or tough connective tissue. The whole girdle can move in the anterio-posterior plane by pivoting on this hinge. Considerable dorsoventral movement is also possible.

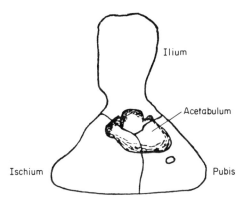

Fig. 6.11 The pelvic girdle of *Eryops*, a rhachitome.

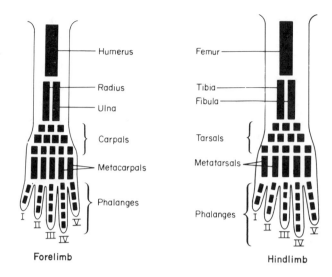

Fig. 6.12 Generalized pentadactyl limbs.

permits rotation of the bone in more than one plane. The distal end of the humerus or femur articulates with two bones the radius and ulna or the tibia and fibula which tend to lie posteriorly. This articulation only permits movement in one plane yielding flexion or extension. At their distal ends these bones each articulate with a small bone, the radiale and ulnare in the forelimb, which themselves articulate with two rows of small interlocking bones, the carpals or tarsals. These form the wrist or ankle where movements are usually largely restricted to a single plane which is not necessarily that in which the radius and ulna, or tibia and fibula, move in relation to the upper limb segment. Finally there are the feet which, in tetrapods generally but not modern amphibians, comprise five short chains of bones each consisting of a metatarsal (metacarpal in the forelimb) followed by the bones of the digits. Modern amphibians have, at most, four digits on the front limbs.

ii Opinions differ about the origins of this organization. The classical and perhaps more widely held view, sees the limbs as derived from the rhipidistian archipterygium. In contrast Westoll has suggested, on the basis of embryological investigations, that this is only true of the proximal components, the archipodium, and that everything distal to the proximal rows of carpal and tarsal bones represents new formations. There are also differences of opinion about the origins of the radius and tibia. Westoll considered that these do not represent bones of the central axis of the archipterygium but rather radials on the anterior, pre-axial, side of the fin.

iii In early amphibians the humerus and femur were held out from the body more or less horizontally. The forearm and foreleg then ran downwards more or less vertically and the hands and feet were turned forwards. The limbs were therefore bent twice; 90° vertically at the elbow and knee, and 90° horizontally at the wrist and ankle.

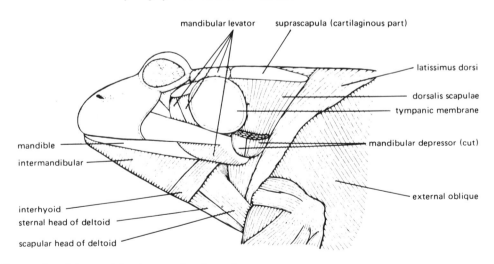

Fig. 6.13 The muscles of the cranial and pectoral regions of a frog. (From Holmes.)

MUSCULAR SYSTEM

(a) Cephalic region

Various components of the musculature are reorganized at metamorphosis. In adult anurans the branchial musculature is reduced (Fig. 6.13). A mandibular levator originating on various parts of the skull persists and is inserted on the hind part of the mandible. This closes the mouth. A mandibular depressor, which opens the mouth, originates from a circular band of tough fibrous tissue to which the edges of the tympanic membrane are attached, and is inserted on the posterior tip of the mandible. A large intermandibularis together with a smaller posterior interhyoideus form a thin sheet on the ventral surface of the head. Their contraction elevates the floor of the buccal cavity thereby aiding breathing and swallowing. A petrohyoideus originating on the hind end of the skull elevates the hyoid apparatus and also assists in swallowing. The cucullaris, also originating on the skull and inserted on the anterior edge of the scapula, elevates and rotates this last. Beneath the intermandibularis lies the geniohyoideus. Acting in concert with sternohyoideus and omohyoideus, this moves the hyoid apparatus.

(b) Trunk musculature

Epaxial muscles lie along the vertebral column from the pelvic girdle to the skull. They are attached to all vertebrae and subdivided into a number of bundles with differing fibre directions but the absence of myosepta obscures the original myotomal pattern. In contrast, the rectus abdominis, which is situated on the ventral part of the body, does bear traces of such septa. Originating on the ventral surface of the pubis, it is inserted on the hind end of the sternum and retracts this. An external oblique muscle is inserted on the sheath of the rectus abdominis, covers the lateral surface, and its contractions compress the body. An underlying transversus originates on the ilium and serves a similar function.

(c) Appendicular muscles

The limbs can be rotated, abducted, adducted and retracted. In the case of the anuran forelimb the latissimus dorsi is inserted on the front of the humeral crest, rotates the limb outwards, and both abducts and retracts it. The deltoideus has two components which protract, adduct and rotate the upper limb. A pectoralis adducts it, and a coracoradialis adducts and protracts it.

The hind limb has comparable muscles that originate from the pelvic girdle. Lower down the limbs flexors and extensors are associated with the various joints.

LOCOMOTION

i The pentadactyl limb is a chain of levers. The proximal joints exert a leverage on the distal ones and these exert leverage against the ground. So long as the foot remains fixed the body is impelled forward by a force equal to that with which the foot presses backwards against the ground. Modern amphibians have a variety of locomotory techniques depending upon the group considered and whether it is moving in water or on land. The Ichthyostegalia and labyrinthodonts were broadly urodele-like and presumably progressed on land as these do today. In the water they can either walk, or close their legs against the body and swim by undulations of the body and tail. On land they use symmetrical,

Fig. 6.14 The motion of a newt when walking.

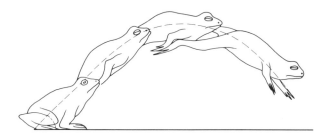

Fig. 6.15 Take-off in saltatorial anurans.

diagonally supported gaits in which the legs move in the order right forefoot, left hindfoot, left forefoot, right hindfoot. This requires energy to hold the body off the ground during movement but the body is supported by the ground at rest. When moving slowly there are always three feet on the ground, the fourth being swung forward. When moving rapidly there are moments when only two feet are on the ground. The body itself curves at each step. The tail has the maximal lateral displacement, followed by mid-body regions (Fig. 6.14), and the amplitude is least in the pectoral and pelvic regions.

ii The walking gaits of anurans are similar to those of urodeles but are slow and the short, tailless body lacks any marked lateral displacements. Such rather cumbersome movements are offset by the ability to jump when the animal propels itself forwards from the sitting position by using both hind legs in unison (Fig. 6.15). Each jump is an isolated event and does not benefit from any residual 'bounce' from the foregoing one as is the case in kangaroos. The single backward thrust generates all the kinetic energy and there are no further sources when the animal is in the air. During push-off the long leg segments, initially folded against each other, unfold more or less synchronously. The longer the legs and the larger the number of folds, the longer the time the body is moved whilst the feet are still pushing against the ground and the farther the body is moved. Complementarily, the lighter the body the smaller the force that is required, and jumping efficiency is also enhanced by streamlining the body during the jump. The origin of the synchronized hind limb movement is contentious and speculative. Some people see it as associated with the movements of swimming, others as a method of quickly returning to the water. Its role in eluding predators is frequently supplemented by the momentary display of flash colouration. This comprises bright colouration, normally concealed at rest when the animals merge with their background as the result of

cryptic colouration, that serves to distract and confuse by its sudden appearance and disappearance.

DIGESTIVE SYSTEM

(a) Buccal cavity

i A well-developed tongue first appears in the Amphibia and is an adaptation to life on land. A fish's food is always wet but that of terrestrial animals has to be moved around the mouth for moistening and chewing. The so-called 'boletoid' tongue of salamanders has an extensive attachment to the jaw but that of most frogs is only attached near to the anterior margin of the jaw and, at rest, is folded back on the floor of the mouth with the tip pointing towards the throat. In both groups mucous glands provide moist secretions.

ii Some larval forms have horny, epidermal, toothlike structures but most adult amphibians have true teeth. These are usually borne on the jaws but may also occur on the palatine and vomer bones. In a few species they occur on the parasphenoids. They are polyphyodont—can be replaced an unlimited number of times—and in many species have the tooth crown separated from the pedicel by a zone of weakness.

(b) Oesophagus and stomach

The adult oesophagus and stomach differentiate at metamorphosis from the ciliated and mucogenic larval foregut. The oesophagus is a muscular tube with sphincter muscles at its end. Short and with a large cross-section in anurans, it is elongate in genera like *Amphiuma*. The mucous epithelium includes one or more layers of cuboidal or columnar cells, many ciliated, with interspersed goblet cells. The adult stomach has a thin wall consisting of a glandular mucosa and well-organized muscular tunic. Lined by slender mucogenic cells it lacks goblet cells and cilia. Chief glands occur in the corpus. They have basal zymogenic cells and large mucous neck cells near to the gastric lumen. Pepsinogen is most abundant in the more anterior ones. Secretion of HCl involves the accumulation of H^+ and Cl^- ions at the gastric surface.

(c) Small intestine

The mid- and hindgut of anuran tadpoles are somewhat longer than the body with maximal relative lengths usually being attained when the hind legs are well developed. The small intestine of the adult anuran is long and folded, that of urodeles runs directly back to the large intestine, and, as in amniotes, it is the principal site of digestion. Anteriorly it is thrown into longitudinal ridges which gradually fuse behind to give crescentic folds enclosing deep crypts. In *Proteus*, *Salamander* and *Necturus* epithelial cells originate in such crypts and migrate to the surface. The entire mucosa is always well vascularized and lymphatic vessels are concentrated beneath the epithelial ridges. The surface consists largely of columnar cells with interspersed mucoid goblet cells and the mucus elaborated by both these and modified Brunner's glands protects the surface from both mechanical and autoproteolytic damage.

(d) The pancreas

The pancreas situated between the duodenum and stomach is of variable form. It has a single secretory duct in anurans, more in urodeles. Secretion is intermittent and involves trypsin, erepsins, amylases and lipases. It is stimulated by the hormone secretin which is produced in the duodenal wall at the time that the acid chyme leaves the stomach.

(e) The liver

Both liver and gall bladder are always present. The liver of gymnophiones and urodeles is undivided, that of anurans is bilobed. Bile is released as the stomach empties and initiates or amplifies peristaltic contractions, thereby mixing the intestinal contents. It prevents bacterial growth, and facilitates the digestion and absorption of fats.

(f) Large intestine

The adult hindgut is usually demarcated from the small intestine by an abrupt widening of the lumen. In *Rana* a flap-like valve prevents retrograde movement of the contents. It has no specific digestive function although the pH of 8.0 allows the processes initiated within the small intestine to continue. Its principal function is as a site for the absorption of water and salts. The semi-solid mass that remains is voided as faeces.

RESPIRATORY SYSTEMS

(a) Gaseous exchange in the larvae

i In most amphibians three pairs of pharyngeal pouches open to the exterior and form gill slits (Fig. 6.16).

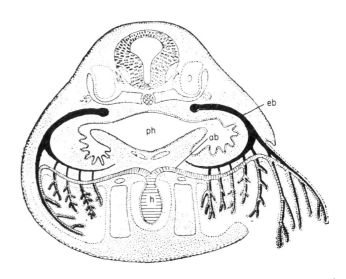

Fig. 6.16 Internal and external gills of a tadpole. (After Orr.) h, heart; ab and eb, afferent and efferent branchial vessels; ph, pharynx.

Water entering at the mouth or nares flows over the gills and out to the exterior. Cartilaginous visceral arches within the intervening septa provide support and the gills themselves consist of filaments covered by ciliated epithelium which are borne by the visceral arches in front of the gill slit. Those of anuran tadpoles are usually smaller and simpler than those of urodeles, and a sheet of tissue, the operculum, soon grows back from the hyoid region to cover both them and the area from which the forelegs will develop. It then fuses with the body wall behind and below the gills providing an atrial cavity which opens to the exterior either ventrally or laterally. Shortly after this the original gills degenerate and new ones develop from the walls of the gill clefts. A small opercular fold also develops in urodeles and caecilians but no atrial chamber is formed. Many larval urodeles have valves around the internal nares that control the direction of water flow.

ii Various specializations occur in particular species. In the viviparous apodan family Typhlonectidae two large baglike gills develop from the dorsal pharyngeal region and probably absorb oxygen from the oviducal fluid. The external gills of many basin-building anurans are very large. Those of the Nototremata, whose young develop in a dorsal brood pouch, are modified and probably absorb nutriment.

iii During metamorphosis the gills are resorbed, the

gill slits close and the lungs, if present, become functional. The gills of the viviparous caecilians are usually resorbed prior to hatching or birth.

(b) Gaseous exchange in the adult

i Adults of the Urodela exhibit more variation in their surfaces for gaseous exchange than do anurans. The family Plethodontidae, although terrestrial, lacks lungs and exchange occurs at the skin, or at the buccal and pharyngeal surfaces. The aquatic genera *Amphiuma* and *Cryptobranchus* develop lungs, lose the gills, but retain one pair of gill slits. Adult Proteidae and Sirenidae possess both gills and lungs. In such forms extirpation of the gills forces the animal to come to the surface and gulp air. Complementarily, if unoperated animals are prevented from reaching the surface their gills enlarge. Experiments which involved separating the air going to the lungs from that going to the body surface showed that more oxygen is taken up at the skin at low temperatures. At high temperatures the pulmonary surfaces take up additional oxygen. Carbon dioxide, which is much more soluble in water, is principally voided through the skin at all temperatures.

ii A complex series of events attends pulmonary ventilation. In frogs the cartilaginous branchial skeleton of the tadpole gives rise, at metamorphosis, to a cartilaginous plate, the corpus of the hyobranchial apparatus. This can be pulled backwards and downwards by the sternohyoideus muscle that originates on the pectoral girdle, thereby enlarging the buccal cavity much as in fishes. Petrohyoideus, geniohyoideus and intermandibularis muscles act to raise the plate and reduce the volume of the cavity.

Closure of the glottis retains air in the lungs under slight pressure. The buccal cavity is enlarged and air drawn in through the nostrils. Opening of the glottis then permits air to be expelled from the lungs into the buccal cavity. The nostrils are then closed, the volume of the buccal cavity reduced, and air is driven into the lungs. Closure of the glottis repeats the cycle. Use of an initial 80%:20% argon/oxygen mixture demonstrated that the pulmonary gas mixture is not exchanged in a single ventilatory cycle, but that some returns from the mouth to the lungs and is only lost after four or so cycles.

URINOGENITAL SYSTEMS

(a) The kidney

Tadpoles excrete their nitrogenous waste as ammonia. Adult frogs convert it to urea. The mesonephric kidneys are long in urodeles and usually have narrow anterior

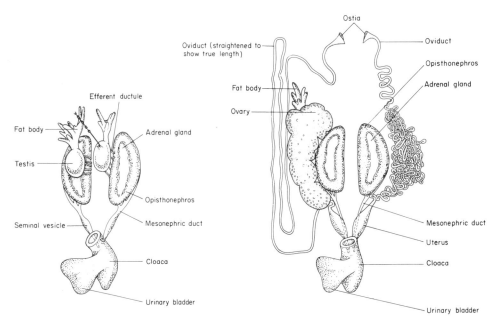

Fig. 6.17 The male and female reproductive systems.

and expanded fully functional, posterior regions. Those of caecilians extend throughout the body cavity and are usually of uniform width throughout. Ranids have a more compact kidney. Caecilians and sirenid urodeles have well-developed glomeruli in the anterior regions but these are reduced or lost in most urodeles and anurans. In general there is a thin-walled urinary bladder which opens directly into the cloaca and lacks any direct connection with the Wolffian ducts from the kidney. Urine passes down these ducts into the cloaca and then backs up into the bladder for storage. The liquid discharged by captured toads is not urine but water stored in the cloaca. It has been suggested that this reduces the weight and facilitates escape.

(b) Male systems

i   The reproductive organs of frogs are represented in Fig. 6.17. They are essentially similar to those of fish although the precise shape of the testes reflects gross body morphology. In some caecilians they are elongated with beadlike swellings consisting of masses of seminiferous tubules connected by a longitudinal duct. Other caecilians have more compact structures. Those of salamanders are shorter and more irregular in outline. Those of frogs are compact. In all cases there is a pronounced difference between the size of the testes and their associated fat bodies in the breeding and non-breeding seasons.

ii   Male bufonids have a peculiar structure, Bidder's organ, between the testes and fat body. Removal of the testes leads to this organ developing into an ovary and the rudimentary oviducts enlarge, presumably under the influence of oestrogens secreted by the newly developed ovaries.

iii   The longitudinal collecting canals of the testis of caecilians send small transverse canals to the ipsilateral kidney. There they join the nephric tubules and enter the Wolffian duct. In salamanders, efferent ductules carry the sperm from the testis to a longitudinal canal, Bidder's canal, on the median edge of the kidney. A number of short ducts connect this with nephric canals in the anterior region of the kidney to join the Wolffian duct. In those genera where the anterior kidney is no longer excretory, the posterior renal ducts open directly into the cloaca and the Wolffian duct is wholly reproductive in function—a true ductus deferens. In some Anura the efferent ductules from the testis connect directly with the kidney ductules. In others they join a Bidder's canal.

Sperm are then transferred through the nephric tubules to the Wolffian duct which emerges near the posterior end of the kidney and passes to the cloaca. In many species they are stored temporarily in a dilation (seminal vesicle) close to the cloaca.

In males of all known urodeles apart from the Hynobiidae and Cryptobranchidae the cloacal lining hypertrophies during the breeding season and secretes a jelly-like substance that unites clusters of sperm within a spermatophore (see p. 90).

(c) Female systems

Amphibian ovaries are saccular, contain a lymphatic cavity, and their shape again reflects that of the body. Those of caecilians are long and narrow; those of urodeles shorter, with a single continuous cavity; and those of frogs compact with the cavity subdivided into compartments. There are again associated fat bodies. These are foliose structures in caecilians and lie along the lateral edge of the ovaries. Those of urodeles are long and slender, and lie parallel to the median edge of the ovary. Those of frogs are anterior to the ovary. In all cases their size decreases as the eggs enlarge and they serve as nutritive stores for these latter. Following release from the ovaries the eggs are wafted towards the ostia of the Mullerian ducts by cilia on the peritoneum, liver, etc. For most of their length these paired ducts have thickened glandular walls that secrete the gelatinous covering of the eggs. A posterior swelling forms a site of temporary storage for the eggs prior to laying. The black colour of the eggs enables them to absorb solar radiation. The external protein gel, which is not eaten by the tadpoles, probably minimizes the loss of this heat, as convection currents, which would occur in neighbouring water if it directly contacted the eggs, are prevented.

ENDOCRINE SYSTEMS

(a) The hypophysis

The adult hypophysis always includes a pars distalis and pars intermedia. A bilobed pars tuberalis is present in anurans but absent from urodeles. The neurohypophysis comprises a pars nervosa and median eminence. Hypophysial portal vessels permit hypothalamic control of the pars distalis. The adenohypophysis contains all the hormones known in other taxa—FSH, LH, prolactin, TSH, ACTH and STH. The pars intermedia secretes melanophore stimulating hormone (MSH)

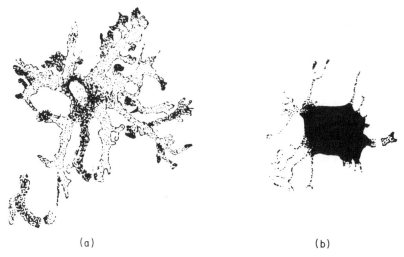

(a)   (b)

Fig. 6.18 Melanophores of a frog dilated (a) and contracted (b). (After Noble.)

which has two different effects on cutaneous melanophores. A rapid effect disperses melanin granules and a slower, more permanent effect, increases the total number present (Fig. 6.18). The neurohypophysis contains the octapeptides mesotocin and arginine vasotocin (see chemical structure at foot of page). These act to counteract dehydration by way of effects upon water movements occurring at the mesonephric kidney, urinary bladder and skin.

(b) The thyroid gland

The thyroid hormones exert well-known and spectacular effects at metamorphosis. Triiodothyronine, although present in far smaller quantities than thyroxin, has a 4–8 times greater effect. This has been interpreted as reflecting a faster penetration of the cells and it has been suggested that thyroxin is converted to triiodothyronine prior to its action. Embryonic tissues are insensitive to the thyroid hormones and their differential sensitivity appears abruptly towards the end of embryonic life. Some tissue changes reflect direct responses whereas others may represent inductive effects exerted by neighbouring structures.

(c) Parathyroid glands

Two pairs of parathyroid glands arise from the ventral region of the third and fourth pharyngeal pouches. In the adult they are situated close to the external jugular vein. Complete parathyroidectomy results in greatly reduced blood calcium levels and high urinary calcium excretion rates. Plasma phosphorus shows a concomitant transitory increase.

Cys. Tyr. Ileu. Glu(NH$_2$). Asp(NH$_2$). Cys. Pro. Ileu. Gly(NH$_2$)

Mesotocin

Cys. Tyr. Ileu. Glu(NH$_2$). Asp(NH$_2$). Cys. Pro. Arg. Gly(NH$_2$)

Arginine vasotocin

(d) Ultimobranchial glands

Paired ultimobranchial glands are derived from the floor of the sixth pharyngeal pouch in anurans at the time that the operculum develops. A single structure occurs in urodeles. They increase in size by the addition of secondary follicles during metamorphosis. This coincides with a period of calcium mobilization from endolymphatic sacs associated with the auditory organs and with posterior prolongations into the vertebral column. These endolymphatic and paravertebral sacs form aragonite stores which are utilized at metamorphosis for ossification of the cartilaginous larval skeleton.

(e) The thymus

The amphibian lymphoid system comprises thymus, spleen and lymph nodes. The thymus arises from the dorsal surface of a branchial cleft. All three organ types filter both blood and lymph and serve as sites of antibody synthesis. Antibody-producing and antibody-carrying cells can be detected by the formation of rosettes with heterologous protein such as sheep erythrocytes.

(f) The adrenal glands

Adrenal tissue occurs in anurans as two strips in the ventral surface of the kidneys. The tissue is not so consolidated in urodeles but is distributed as islets on the ventral surface of, or within, the kidneys. At least three types of cells have been described. One type is the homologue of the cells in the mammalian adrenal medulla and secretes adrenalin and noradrenalin. A second type corresponds to the amniote cortex and induces shifts in water and mineral metabolism, probably by affecting membrane permeability in muscle and other tissues. A third type consists of the summer cells of Stilling which may contain glycoprotein material. They have no obvious homologues in amniotes.

(g) The endocrine pancreas

The islets of Langerhans are widely scattered within the pancreas and comprise branching cords of cells. They always produce insulin but it is not clear whether they have the uniform differentiation into α and β cell types as in birds and mammals. One cell type has been described in newts of the genus *Triturus*, two in many amphibians, and five in *Rana temporaria*.

CARDIOVASCULAR SYSTEM

(a) Arterial system

i The overall pattern of the cardiovascular system retains the significant features detailed for the various taxa of fishes. In all larval amphibians, and in the adults of perennibranchiate urodeles, the aortic arches, instead of being interrupted by a capillary anastomosis as happens in fishes, give off vascular loops which pass into the gills and there branch into capillaries. These then reassemble and rejoin the aortic arches dorsally. During metamorphosis the vascular loops degenerate, the arches expand, and the flow of blood from heart to body continues without interruption.

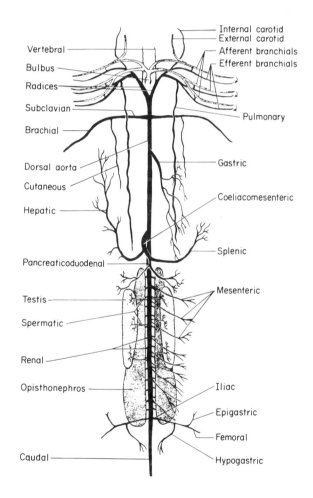

Fig. 6.19 The arterial system of *Necturus*. (After Weichert.)

ii Urodeles (Fig. 6.19) exhibit fewer differences than frogs when compared with the ancestral condition. The first two arches are lacking and the forward extensions of the ventral aorta from which they arose run forward to the jaws as the external carotid arteries. The third pair of arches form the internal carotids supplying the face and brain. A connection frequently persists with the fourth pair of arches, the systemic arches, which give rise to the main parts of the dorsal aorta. The fifth pair can persist, although may be reduced in size. The sixth pair give the pulmo-cutaneous arteries which carry blood, mainly from the right side of the heart, to the skin and lungs where oxygenation occurs. The ductus arteriosus, a residual connection with the dorsal aorta, persists on each side and enables some of the blood in the sixth arch to pass to the body rather than to the sites of gaseous exchange.

iii The arterial systems of most adult anurans differ from this in a number of ways. No connection ever exists between the third and fourth arches; the fifth pair disappear, and the sixth lose their connection with the dorsal aorta via the ductus arteriosus. Unoxygenated blood is, therefore, all channelled to the lungs or skin.

(b) The heart

i The anuran heart has separate left and right atria but an undivided ventricle. Blood from the lungs returns to the left atrium, that from the rest of the body to the right one. This latter, therefore, contains a mixture of blood that has been oxygenated in cutaneous regions, and blood that has relinquished its oxygen to the tissues. The ventricle has rather spongy walls that probably reduce swirling and, therefore, minimize mixing of the blood from the two atria. The truncus arteriosus is partially divided by a spiral valve with the pulmo-cutaneous arteries opening from one side of the partition, the systemic and carotid arches from the other. As in the case of lungfish, this leads the oxygenated blood, that has recently traversed the lungs, to the carotid and systemic arches, and hence to the general body circulation. The remainder tends to be directed to the pulmocutaneous regions. Experimental investigations have shown that some mixing certainly occurs in the ventricle but that this is far from total.

ii Urodelan hearts differ from the foregoing as they are less compact and the truncus arteriosus is simpler. Terrestrial genera possess a spiral valve but the horizontal septum of the truncus arteriosus is not so developed.

The lumen of the truncus is subdivided into four channels corresponding to the four pairs of persistent arches. Spiral valves tend to be absent from perennibranchiate genera in which the two streams of blood are more confluent and a similar situation characterizes gymnophiones. In lungless salamanders the absence of any division in the atrial region can give an essentially fishlike condition.

(c) The venous system

A venous system is represented in Fig. 6.20. In all essential features it resembles those described for fishes. The veins from the intestinal region join to form a hepatic portal vein that branches into sinusoids in the liver. The blood is then gathered into hepatic veins and returned to the heart. Blood returning from the hind legs

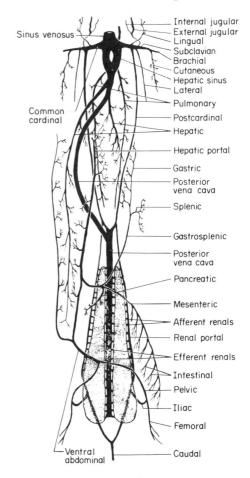

Fig. 6.20 The venous system of *Necturus*. (After Weichert.)

and caudal regions is emptied in part into the hepatic portal vein through an anterior abdominal vein, whilst the rest passes through the renal portal system. Blood from the lungs returns to the heart via pulmonary veins.

SENSE ORGANS

(a) Olfactory organs

i Olfactory sensilla occur in the epithelium lining the passages linking the external and internal nares. In aquatic amphibians, as in crossopterygian fishes, currents of water pass through this passage. In terrestrial adults, as in amniotes, these passages are filled with air and the epithelium covered by mucus. Currents then pass as a result of the ventilatory cycles. In both terrestrial and aquatic phases or genera the currents are controlled by valves, at the inner end in many aquatic genera and at the outer end in many Anura. Some Anura are able to close the nostrils because tubercles on the anterior end of the lower jaws move forward as the mouth is closed and push the premaxillary bones apart.

ii Each olfactory sense cell bears numerous cilia of varying length. These project into the water in aquatic forms and the longest penetrate right through the overlying mucus and into the air in terrestrial forms. In the larvae some cilia beat but some at least are presumed to be effective receptors.

iii A specialization that first appears in Amphibia is Jacobson's organ. This is a patch of modified epithelium situated on each side of the wall of the nasal passages and sometimes sunk into a diverticulum. It is most highly developed in squamate reptiles.

(b) Taste

Taste buds are limited to the buccal cavity and occur on the jaws, on the palate and on the tongue where they are localized on fungiform papillae. When stimulated they elicit snapping or swallowing. A more diffuse chemical sensitivity occurs over the skin.

(c) Cutaneous sensory organs

Various tangoreceptors and chemoreceptors occur in the skin. Lateral line organs occur in both aquatic larvae and the adults of many perennibranchiate genera but disappear under the skin during terrestrial periods in salamanders. They are simpler than those of fish, comprise groups of epidermal sense cells, usually in pits, that are sometimes scattered and sometimes organized in rows. The pits are not joined by canals in adults of extant genera. The tactile, general chemical and gustatory systems are all intimately related centrally. General chemical sensibility persists if the taste buds are denervated but the specific gustatory function of the taste buds seems to disappear if the general cutaneous innervation is eliminated!

(d) Photoreceptors

i LATERAL EYES
The majority of amphibians have well-developed eyes although they are somewhat degenerate amongst secondarily aquatic anurans, most caecilians and various cave-dwelling urodeles. Life on land exposed the delicate, damp, external surfaces to the air, made it possible to see further than is possible through water, but increased the refraction that occurs at the interface between the cornea and the external medium. Aquatic animals suffer very little such refraction. Eyelids occur in all tetrapods. Many fish have immovable 'adipose' or vertical lids but only some sharks, such as *Galeorhinus*, have ventral and dorsal lids comparable with those in terrestrial vertebrates. These are assumed to represent convergent evolution. Harderian glands secreting an oily material first appear in amphibians, as do lachrymal ducts connecting the space within the lower lid to the nasal cavity. It has been suggested that the latter are derived from a component of the external nares. The Anura also have a nictitating membrane—possibly a parallel evolution to those in amniotes. In all cases it is a thin sheet formed from the rim of the ventral lid whose edge is drawn out as a strand around the eyeball. In Anura it passes through the retractor bulbi muscle behind the eyeball. This is a modified part of the superior rectus muscle and its main function is to draw the eyeball back into the head but when it contracts it also draws the nictitating membrane over the cornea.

The lens is hard and almost spherical (Fig. 6.21). The resting eye is focused on near objects and accommodation is achieved by protraction of the lens. This contrasts with the situation in other tetrapods where the resting eye is focused for distant vision and is actively accommodated for near vision by deformation of the lens. The spherical lens also gives more refraction than the ellipsoid ones of amniotes.

The retinal systems include receptor, bipolar and ganglion cells. The receptors include rods, cones and double cones to a total of c.1×10^6. There are 2.5–

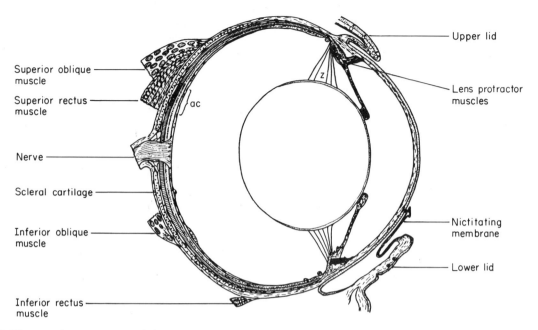

Fig. 6.21 The eye of an anuran amphibian in vertical section. (After Walls.) z, zonule; ac, area centralis.

3.5×10^6 connecting cells and 0.5×10^6 ganglion cells. Ultrastructural studies on *Necturus* demonstrate clear synaptic relationships (Fig. 6.22). The outer segments of the receptors undergo constant renewal.

ii THE PINEAL COMPLEX

The pineal complex of anurans is a functional photoreceptor including an end vesicle in the skin and a tubelike pineal body connected with the brain (Fig. 6.23). Caecilians and urodeles only possess an intracranial organ. The end vesicle exhibits two types of response. Achromatic responses involve the depression of impulse activity to all wavelengths; chromatic responses involve increased impulse frequency in response to medium and long wavelengths, or a decreased frequency in response to short wavelengths. Electrical activity in the intracranial pineal requires neither the presence of the lateral eyes nor that of the end vesicle. The presence of serotonin and its known effects on chromatophore dispersion suggest that one function is to control body colour. The presence of hydroxyindole-o-methyl transferase (HIOMT) in both the retina and pineal complex compares with the situation in fishes but the retinal levels are lower than in reptiles and birds. It is intimately involved in the synthesis of melatonin which

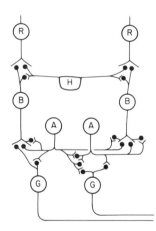

Fig. 6.22 The interrelationships of retinal cells in *Necturus*. H, horizontal cell; B, bipolar cells; A, amacrine cells; G, ganglion cells; R, receptor cells.

contributes to the control of circadian and circannian cycles.

(e) The ear and hearing

i Many anurans produce distinct species-specific vocalizations which are necessarily associated with marked

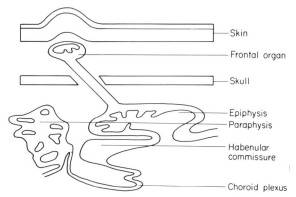

Fig. 6.23 Diagram showing the relationships within the pineal complex of a frog. (After Wurtman *et al.*)

Labels: Skin, Frontal organ, Skull, Epiphysis, Paraphysis, Habenular commissure, Choroid plexus

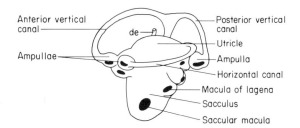

Fig. 6.24 The inner ear of a frog. De, ductus endolymphaticus. (After Retzius.)

Labels: Anterior vertical canal, Ampullae, de, Posterior vertical canal, Utricle, Ampulla, Horizontal canal, Macula of lagena, Sacculus, Saccular macula

hearing abilities. Terrestrial life has actually resulted in greater changes to the ear than to the eye giving the increased sensitivity necessary for efficient hearing in air, where the vibrations impinging on the ear tend to be reflected more than in the denser aquatic medium. The inner ear is similar in both larvae and adults but the middle and external ears are very different.

ii Tadpoles use their air-filled lungs as eardrums. The anterior portion of each lung is connected to the ipsilateral round window by way of a bronchial columella. Slight changes in water pressure cause variations of lung volume and the columella transmits these to the inner ear via the round window. The oval window acts as a pressure release mechanism. During metamorphosis the bronchial columellae disintegrate and there is a simultaneous development of an external tympanic membrane. This last transforms sound pressure into a displacement of a cartilaginous tympanic columella in the middle ear which, in turn, produces a mechanical deformation of the membrane of the oval window. The membrane of the round window now acts as a pressure release system so that the roles are reversed. Another pressure release system involves the eustachian tube which links the air filled cavity of the middle ear to the pharynx.

iii As the diameter of the tympanic membrane is greater than that of the oval window their size difference assists in matching the acoustic impedance of air to the higher impedance of the fluids in the inner ear. The membranous labyrinth of the inner ear consists of the well-known endolymphatic and perilymphatic fluid systems (Fig. 6.24). The vestibular receptor organs, sacculus, utriculus, lagena and semicircular canals, together with two auditory organs, the amphibian papilla and the basilar papilla, are situated in the endolymphatic system. The arrangement of these two auditory papillae varies in different genera and species. Indeed in various anuran genera that inhabit cascading streams, shallow water or deep lakes, such as *Ascaphus, Pipa* or *Telmatobius*, hearing is reduced or lost.

iv The basilar papilla is a simple organ with 50–60 hair cells, located in a short tube which opens to the endolymph system at one end and is separated from the perilymph system at the other end by a contact membrane. Each hair cell bears a number of stereocilia and one kinocilium which is on that side of the cell nearest to the contact membrane. The cilia are all attached to the overlying tectorial membrane. The amphibian papilla is more complex. Its sensory epithelium is twisted and it contains 600 hair cells that are similar to those of the foregoing. The tips of the stereocilia are embedded in a gelatinous tectorial membrane and the orientation of the kinocilia varies in a complicated manner.

AUTONOMIC SYSTEM

The autonomic system comprises cranio-sacral parasympathetic, and thoracico-lumbar sympathetic divisions. Both components include peripheral ganglia. The parasympathetic elements include the autonomic fibres in cranial nerves III, VII, IX and X. Sympathetic fibres run from the cells of origin in the spinal cord through *rami communicantes* to their segmental ganglia. These ganglia are united into a chain by longitudinal fibre bundles, the *rami interganglionares*. Post-ganglionic fibres run to the glands or smooth musculature. One additional sympathetic component originates in the intracranial prootic ganglion. A great variety of visceral controls are exerted on the intestinal tract, the renal region and the reproductive system.

THE BRAIN

(a) The medulla oblongata

The medullary region of tadpoles and perennibranchiate urodeles resembles that in fishes. The whole brain also resembles that of teleosts in its ability to regenerate following damage. At metamorphosis major changes occur in the medulla and, in particular, the extensive dorsal nuclei which are associated with the lateral line system degenerate so that only small anterior vestibular and auditory nuclei occur in the adult. The motor and sensory nuclei are generally embedded in the extensive brain stem reticular formation and retain an overall longitudinal disposition reminiscent of that in selachians.

(b) The cerebellum

The cerebellum of perennibranchiate urodeles remains relatively undeveloped. Elsewhere a bipartite central corpus is flanked by auricular lobes.

(c) The mesencephalon

The optic tectum forms a prominent dorsal component whilst a well-developed auditory nucleus, the torus semicircularis, is a major relay station on the ascending auditory pathway of each side. The tectum integrates impulses of tactile, proprioceptive, visual, and, in the larvae, lateral line, origin. Descending effector systems include a prominent tectobulbar pathway.

(d) The diencephalon

The epithalamic region includes the habenular nuclei and associated pineal complex (q.v.). The left habenular nucleus is bipartite and such asymmetries occur in various classes. It has been suggested that they reflect the differential prominence of ancestral pineal and parapineal units. The predominant diencephalic components are, however, in the thalamus whose numerous component relay nuclei are arranged in rostro-caudal layers. The magnocellular preoptic nucleus in the hypothalamus is connected to the neurohypophysis, and distinct parts of the small-celled tuberoinfundibular nuclei are connected, via the median eminence, with hypophysial portal vessels. These pathways mediate the hypothalamic controls over hypophysial secretion.

(e) Cerebral hemispheres

Those of urodeles are longer than those of anurans. During development, dorsal pallial and ventral basal zones can be detected. In the adults the first-named of these gives rise to the primordial hippocampal, general pallial and primordial pyriform regions. The basal zone gives rise to medial septal and more lateral paleostriatal areas. The degree of differentiation of the striatal regions appears to closely parallel the relative development of Jacobson's organ since this is large in anurans and caecilians which have well-developed paleostriatal and archistriatal regions.

ADAPTIVE RADIATION

(a) Urodela

i The modern urodeles are quite diverse. One family that is prominently represented in Europe is the Salamandridae to which the newt genus *Triturus* belongs. All have a more or less elaborate courtship. For example, breeding in *Euproctus asper* takes place in the water. The male approaches the female, pushes his body under hers, and, at the same time, twists his tail around her pelvic region. The female is unable to free herself and this period of amplexus ends when a spermatophore is transferred from the male to the female cloaca.

ii Another family is the Plethodontidae. These are the lungless salamanders that inhabit a wide range of habitats in America. One species occurs in southern Europe. The Cryptobranchidae and Proteidae are other families that retain their gills throughout life. The largest living urodele, *Megalobatrachus*, belongs to the Cryptobranchidae. The proteids are mostly cave dwellers with rudimentary eyes and elongated bodies; the Hynobiidae includes some two dozen species mostly living in Japan and south east Asia, and the Sirenidae comprises two genera in south eastern North America. The maximal extent of the ice sheets during the Quaternary glaciations had a very great impact on the distribution of urodeles, which were restricted to relict populations in ice-free areas.

(b) Anura

i The adaptive radiation of Anura is exemplified by the summary diagram (Fig. 6.25) which shows the principal modes of life. There are no cursorial anurans, speed is attained by jumping, and species such as *Rana oxyrhyn-*

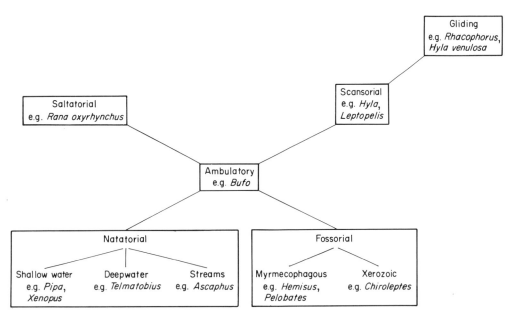

Fig. 6.25 Diagram representing the adaptive radiation of the Anura.

ehus have very long hind legs. Fossorial behaviour can be associated with searching for ants, with preventing desiccation, or with protection from enemies. General like *Hemisus* lack eyelids and have degenerate eyes, a pointed snout and very powerful shovel-like hind legs. *Chiroleptes* lives in regions where there is sparse rain and absorbs water during the brief rainy season. Fully aquatic genera can inhabit shallow stagnant water, *Pipa americanum*; deep lakes, *Telmatobius*; or mountain streams, *Ascaphus*. *Pipa* lacks both eyelids and tympanic membranes. *Telmatobius* has a reduced skeleton and lacks lungs, whilst *Ascaphus* has a reduced middle ear, reduced lungs and practises internal fertilization by cloacal extrusion.

ii Many genera have specialized reproductive behaviour. Thus, various hylid tree-frogs lay their eggs in basins walled-off from streams with mud by the males. Some close relatives use the water in hollow trees or *Bromelia* leaves. In such cases the young are frequently cannibalistic. Genera such as *Polypedates*, *Chiromantis* and *Phyllomedusa* use their oviducal secretions to make foam nests on leaves. The emerging tadpoles then fall into streams below. Brood care occurs, for example, in *Phyllobates*, *Hyla evansi*, *H. goeldi* and *Nototrema*. Male phyllobatids carry the young attached to them by suckerlike mouths. Females

of *H. evansi* carry them on their backs and in *H. goeldi* there is a dorsal depression in which they remain. In *Nototrema* skin grows over such a depression to enclose the young in a brood pouch. Various parts of the alimentary tract are also used for similar purposes. Buccal incubation occurs in *Hylambates*, with the female undergoing temporary starvation, and the singing pouch is used in male *Rhinobates*. Direct development with an abbreviated life history typifies *Eleutherodactylus* whose young lack both gills and gill slits and develop within foam nests to emerge as young frogs in 21 days. Ovoviviparity is rare but occurs in *Nectophrynoides vivipara* where the embryonic tail is adapted for absorbing food.

FURTHER READING

GOIN C.J. & GOIN O.B. (1971) *Introduction to herpetology.* W.H. Freeman & Co., New York. (N.B. In the UK, herpetology refers only to reptiles. This and Porter include sections on Amphibia.)

MOORE J.A. (ed.) (1964) *The Physiology of Amphibia.* Academic Press, London. (Later volumes edited by Lofts B.)

PORTER K.R. (1972) *Herpetology.* W.B. Saunders and Co., Philadelphia.

SMITH M.W. (1951) *The British Amphibians and reptiles.* Collins, London.

7 · Class Reptilia

Synopsis

Class Reptilia

Sub-class 1 Anapsida—*Cotylosauria, Chelonia
       *2 Synapsida—Pelycosauria, Therapsida
      *3 Ichthyopterygia—Ichthyosauria
      4 Lepidosauria—*Eosuchia, Rhyncho-cephalia, Squamata.
      5 Archosauria—*Thecodontia, *dino-saurs, *pterodactyls, Crocodilia.
     *6 Synaptosauria—Plesiosaurs.

INTRODUCTION

The class Reptilia includes all the poikilothermal amniotes and may also include some homoiothermal forms amongst its fossil genera. They are more fully adapted for life on land than are the amphibians and, in particular, lay cleidoic eggs (q.v.). Although it is possible that all of them have a common origin amongst predecessors of the known gephyrostegid seymouriamorphs, they may, equally, reflect polyphyletic attainment of amniote characteristics. They dominated terrestrial faunas during the Permian, Triassic, Jurassic and Cretaceous, and several lineages, of which the turtles, ichthyosaurs and plesiosaurs are the best-known, returned to a secondarily aquatic way of life. Modern forms comprise the Chelonia, the turtles and tortoises; the Rhynchocephalia represented by *Sphenodon*, the Tuatara of New Zealand; the Squamata, or lizards and snakes, and the Crocodylia.

INTEGUMENT

i The skin of reptiles is not an important site of gaseous exchange and is not moist. The main problem such animals face is indeed one of water loss and although the scale-covered skin is not wholly impermeable many lizards and snakes live in desert conditions.

ii The epidermis has a well-developed stratum corneum, which is shed periodically, and the dermis contains many chromatophores and is both thick and soft in crocodiles. Very few epidermal glands are present and those that are frequently produce pheromones. Crocodiles have musk-secreting glands on the lower jaw and around the cloaca, whilst some turtles produce similar sex attractant substances both in the jaw region and along the line demarcating the ventral plastron from the carapace.

iii The scales are variously of epidermal or dermal origin. The epidermal ones are of two types. Those of lizards and snakes are formed by the cornification of thickened areas of the integument and are continuous at their bases (Fig. 7.1). New sets develop beneath the old ones and these are then shed, lizards creeping out of the old skin or shedding it in flakes, whilst snakes habitually turn it inside out. The rattles of rattlesnakes are modified scales that are not shed with the rest of the skin. In contrast, the epidermal scales of crocodiles and chelonians develop as individual structures. Furthermore, those of tortoises and turtles do not correspond to the dermal plates of the shell, from which they are separated by a thin germinative layer.

iv Dermal scales underlie the epidermal ones on the throat of crocodiles. The ventral plastron of chelonian shells is also formed by fusion of such dermal scales, and the dorsal carapace by fusion of dermal scales, ribs and vertebrae.

SKELETON

(a) The skull

i The skull tends to be higher and narrower than is the case in amphibians and many roofing elements disappear. In particular, there is either a reduction, or a loss, of supratemporals, tabulars and postparietals. The skull of most groups is also variously fenestrated in the lateral or cheek regions. The temporalis muscles that close the jaws originate in these regions, an edge for their attachment is important, and their action is apparently facilitated if an opening develops through which they may bulge when contracting. Cotylosaurs and modern

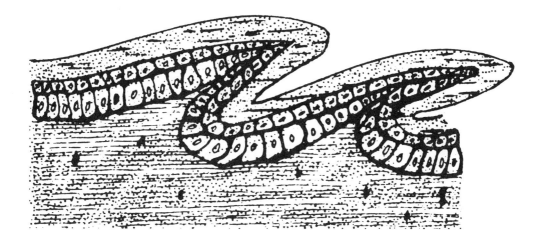

Fig. 7.1 Diagram showing the structure of reptilian scales.

Chelonia lack any such fenestration but amongst the remaining taxa there can be either one or two fenestrations on each side. Each gives rise to a temporal fossa or vacuity. Mammal-like reptiles have one such vacuity that is situated low on the skull (Fig. 7.2) and bounded above by the squamosal and postorbital bones, this is known as the synapsid condition. Protorosaurs and plesiosaurs also had a single vacuity on each side but in these cases it was situated high up on the skull and above a broad cheek that was largely formed by the squamosal. This is known as the parapsid condition. An upper vacuity also characterized ichthyosaurs but here the lateral border was formed by the postfrontal and supratemporal bones. Lepidosaurs and archosaurs such as dinosaurs and crocodiles have both upper and lower vacuities, the diapsid condition (Fig. 7.3).

ii Cotylosaurs were uniformly anapsid, although the diadectomorphs, of somewhat controversial and possibly amphibian relationships, had highly modified skulls. Short and massive, they had prominent supratemporals and both tabulars and postparietals were better developed than is the case in most reptiles. Posteriorly a deep excavation in the cheek region accommodated the ear-drum, and powerful quadrate bones joining the jaws to the braincase were exposed in the borders of these otic notches.

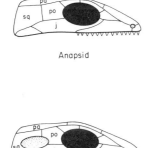

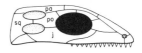

Fig. 7.2 Stylized drawings of the skulls of an anapsid and a synapsid reptile showing the relationships of the parietal, postorbital, squamosal and jugal bones, and the lower temporal opening of synapsids.

Fig. 7.3 Stylized diagrams of parapsid and diapsid skulls showing the superior temporal opening of the former and the two openings of the latter (cf. Fig. 7.2).

iii Modern tortoises and turtles also have highly specialized skulls. They have no vacuity but the skull is excavated to accommodate the jaw muscles. The palatal bones have fused to give a solid structure and, at the side, the prootic and opisthotic bones fuse with the quadrate and squamosal so that the region of the jaw articulation is firmly bound to the braincase. The pineal eye no longer gains access to the exterior of the skull roof; the two external nostrils have a common opening; tabulars, postparietals, supratemporals and intertemporals are lost and a single element, usually interpreted as a prefrontal, occupies the position of this bone itself, together with that of the nasal and lachrymal.

iv As noted above, the diagnostic feature of synapsids was the single lateral temporal vacuity. Small in early genera it became enlarged in many forms and could then extend upwards to reach the parietal. As in most reptiles the ear-drum was situated posteriorly, and above the jaw articulation, but a definitive otic notch was seldom present. Some may have depended on ground-borne vibrations. Although fused postparietals were frequently displaced, along with the tabulars, on to the back of the skull the majority of bones were retained although some elements in the orbital region could be reduced in number.

v Amongst parapsids the genus *Araeoscelis* retained many primitive features. The derivative plesiosaurs retained a pineal eye and there was no secondary palate but the interpterygoid vacuities were nearly or fully closed by fusion of the pterygoids.

vi *Youngina*, an early diapsid, also had a comparatively primitive skull. The palate was generalized, the pterygoids were movable, and small supratemporal, postparietal and tabular bones were present at the back of the skull. *Sphenodon* retains a primitive skull but amongst the Squamata the lizards have their cheek region open behind the eye. The quadratojugal has disappeared, and the squamosal is reduced, so that, as a result, the quadrate is freely movable. The lachrymal is small or absent and the post-parietal is lost, as is either the tabular or supratemporal since a single bone represents both. On the other hand some genera have new superficial bones that cover the skull roof.

vii A rather special condition occurs in snakes, which arose during evolution from a lizard stock. They have skulls which are highly modified for swallowing prey and here even the upper arch of the cheek is lost so that the quadrate is only loosely attached to the skull and palate. The anterior part of the skull can also move freely on the braincase, the dentary is relatively free from the more posterior jaw elements, and the jaws of the two sides are connected by ligaments alone. Swallowing of large prey is facilitated by dislocation of the jaws and movement of the quadrate relative to the upper jaw—streptostyly. Parietals and frontals have also grown downwards to form a new area of ossification that encloses the front of the braincase and protects the brain during swallowing.

viii The remaining diapsid genera comprise the archosaurs—crocodiles, dinosaurs and pterosaurs. The early archosaurs had an antorbital fenestration between the orbit and nostril that lightened the skull, the quadrates were large and the pterygoids met to form a median plate in the roof of the mouth. They include both birds and pterosaurs amongst their descendants. The pterosaur skull was rather archosaurian in appearance but, as in birds, the bones tended to be fused together. The quadrates slanted forward so that the jaw articulation tended to lie below the orbits, which were large, and nostrils were situated on a long beak.

ix Dinosaurs fall into two principal groups. The first of these, the Saurischia, includes the largest tetrapods of all time but they had absurdly small skulls. In contrast, the second group, the Ornithischia, had long, low skulls of a heavier construction and, in various genera, bearing rather bizarre specializations (Fig. 7.4). The horned skull of *Triceratops* made up a third of the body length, whilst the hadrosaurs had duckbills and, sometimes, marked hollow prominences. Amongst these last-named forms the nasal and premaxillary bones formed a crest and its internal cavity was connected with that of the nostrils. This was most elaborate in the genus *Parasaurolophus* where a spinous process projected backward from the skull and may have been used as a snorkel when the animal was underwater (Fig. 7.22).

Fig. 7.4 Reconstruction of the ornithischian genus *Triceratops*.

(b) The axial skeleton

i A typical reptilian vertebra comprises a ventral spool-shaped centrum and, rising above it, a neural arch that encloses the spinal canal. The centra tended to become longer during evolution. Primitively they were amphicoelous, with a concavity at each end, but the ends may be flat, a condition referred to as platycoelous, or successive vertebrae may fit together by virtue of one end being hollowed out and the other one swollen to fit such a socket. Depending upon whether the concavity is in front or behind they are then referred to as procoelous or opisthocoelous. In Jurassic and Cretaceous saurischians the opisthocoelous condition was common and there were ten elongate cervicals, thirteen trunk and four sacral vertebrae. These last were fused to the ilium. In Chelonia there are only ten trunk vertebrae and these are immovably attached to the median plates of the shell. One interesting feature of the tail region in lizards is its tendency to break off at a particular point. This is known as autotomy and facilitates escape.

ii Ribs also vary. The points of attachment of two-headed ribs may shift, or a single-headed articulation with the vertebra may develop. In addition reptiles can have other bony splint-like structures, known as abdominal ribs, ranged along the venter.

(c) Girdles

i The pectoral girdle frequently possesses coracoid, precoracoid and scapula components accompanied by clavicles and interclavicles. Amongst the Chelonia, where dermal elements are incorporated into the plastron, the scapula includes a slim ascending blade plus a long projection to the clavicles. As the coracoid contributes a third prong, the whole structure has a triradiate appearance. In these animals both the pectoral and pelvic girdles are also aberrantly situated as they lie inside the ribs rather than outside them. Amongst the aquatic plesiosaurs and ichthyosaurs, expanded ventral elements formed extensive origins for the large swimming muscles.

ii In the chelonian pelvic girdle a major part of the ventral region is occupied by a large fenestration that almost completely separates the pubis and ischium. Amongst ichthyosaurs, connections between the girdle and sacral vertebrae were lost, whilst dinosaurs exhibited a characteristic series of modifications. In saurischians (Fig. 7.5) the pubis and ischium, no longer

platelike, were directed ventrally and, together with the ilium, gave rise to a triradiate pelvis. Amongst ornithischians (Fig. 7.6) the pelvis had a tetraradiate appearance.

(d) Limbs

i The differing methods of terrestrial locomotion, the ability to fly, the reconquest of the sea and the huge size of many genera are reflected in the great variation of the

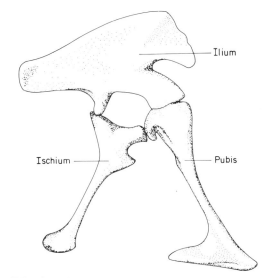

Fig. 7.5 The pelvic girdle of the saurischian genus *Allosaurus*. (After Gilmore.)

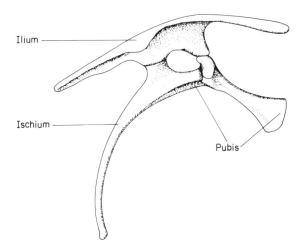

Fig. 7.6 The pelvic girdle of the ornithischian genus *Monoclonius*. (After Brown.)

limbs. The carpus and tarsus are frequently reduced but the number of joints in the toes is fairly uniform. Although many orders developed a more cursorial quadrupedal gait than that of early forms the saurischian dinosaurs passed through a definitive bipedal phase. In these animals the body tilted forward from the pelvis and was balanced by the tail. This was associated with various modifications to the limb bones. The entire weight was supported from the hips, the acetabulum was located high up near the vertebrae, and the hind legs were turned forward. They comprised two powerful pillars and underwent a fore and aft motion (see further on page 97).

ii The limbs of ichthyosaurs (Fig. 7.7) were reduced to short, steering paddles as the main propulsive thrust was generated by undulations of the trunk and tail, and increases in the number of digits, hyperdactyly, or joints, hyperphalangy, were common. In contrast, plesiosaurs had long humeri and femora and the more distal components were shortened. They too exhibited hyperphalangy but there was no hyperdactyly.

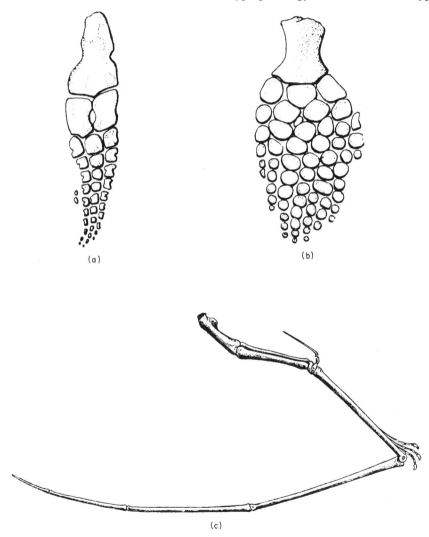

(a)

(b)

(c)

Fig. 7.7 (a) and (b) contrasting ichthyosaur limbs. (c) A pterosaur wing.

iii Pterosaur limbs were also very modified. The humerus was short, powerful, and usually bore a prominent process for the insertion of the breast muscles. The radius and ulna were long and closely apposed whilst the carpus was short, rather rigid, and bore a splintlike pteroid bone (Fig. 7.7) at its anterior edge. This last seems to have supported a flap of skin running up onto the neck. No traces of a fifth digit occur and the four metacarpals were closely associated on their path to the base of the fingers. Three of these fingers were short, though equipped with the normal number of joints and bearing claws. The fourth digit was greatly elongated and its four long phalanges supported the wing membrane which spread between it and the side of the body.

The pelvic limbs were also different from those in other reptiles. Long but slim it would have been difficult for the animals to walk normally on them.

MUSCULAR SYSTEM

i The muscular system is also more adapted to the demands of a terrestrial life than is that of amphibians. This applies particularly to the region of the vertebral column where the musculature assists in the support of the body weight. In most genera few traces remain of the segmental arrangement in fishes, whilst enlargement of the limbs and their muscles further breaks up the continuity along the body. In the tail region epaxial elements continue past the girdle and hypaxial elements preserve a segmental pattern.

ii Epaxial components of the lateral musculature are less modified than are the hypaxial ones and contribute to the longissimus dorsi, which lies alongside the vertebral column, and to the many small transverso-spinal muscles between the vertebrae themselves. The iliocostalis on the dorsal flank is also of epaxial origin. The longissimus dorsi control vertical movements of the vertebral column and are opposed by a subvertebralis complex of hypaxial origin.

iii Hypaxial elements give rise to the intercostals that move the ribs during respiration, and to the rectus abdominis, transverse and oblique muscles that form the body wall and support the viscera. Additional hypaxial derivatives, such as the serratus and the levator scapulae that pass from the scapula to the ribs, suspend the body in the shoulder region. In the ventral part of the neck hypobranchial derivatives join the hyoid and larynx to the pectoral girdle.

iv Neck muscles such as the trapezius and sphincter colli are of branchial origin. The last-named is homologous with the amphibian superficial constrictor and this also contributes to the depressor mandibuli that is attached to the hind tip of the jaws and opens them. The temporalis that closes the jaws is homologous with the adductor mandibulae of fishes.

v Movement of the forelimbs is achieved in part by the deltoideus that originates on the scapula and clavicle, is inserted on the humerus, and pulls this bone forward. A scapulohumeralis assists in holding the limb up. Ventral muscles that are inserted on the humerus include the pectoralis, originating on the sternum and ribs, and the supracoracoideus which originates on the clavicle and interclavicle. Movement of the hind limb involves a levator, the iliofemoralis, an internal pubo-ischio-femoralis that moves it forwards, and a caudifemoralis that moves it backwards.

LOCOMOTION

i Chelonians possibly possess terrestrial locomotory patterns which most closely approximate to the ancestral form, although movement of the axial skeleton is prevented by the shell. Crocodiles have three gaits when on land. In the common, slow gait the body is carried close to the ground and the upper limb segment is held close to the horizontal. In the fast walk the body is raised higher, and in the gallop of small individuals the fore and hind pairs of legs move more or less in unison, the former for propulsion and the latter for support.

ii The powerful hindlegs of lizards permit transitory bipedalism analogous to that of saurischians. A bipedal gait can be more economical of energy as the length of stride can be increased and the forelegs do not persistently act as brakes when they are touching the ground. It is characteristically found in genera of open habitats, or genera which are semi-arboreal, or semi-aquatic and living near forest streams.

DIGESTIVE SYSTEM

(a) Buccal cavity

i The buccal cavity is armed with teeth and equipped with various glands. Early fossils had conical teeth, and piscivorous forms, such as the fossil ichthyosaurs, plesiosaurs and nothosaurs, had a large number. Carnivorous genera such as the carnosaurs had laterally

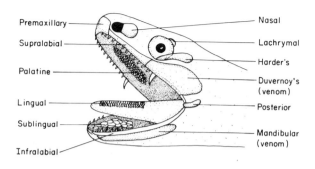

Premaxillary — Nasal
Supralabial — Lachrymal
— Harder's
Palatine — Duvernoy's (venom)
Lingual — Posterior
Sublingual — Mandibular (venom)
Infralabial —

Fig. 7.8 Stylized diagram of a hypothetical reptile showing principal oral glands that exist in different orders. (After Kochva.)

compressed teeth which were often equipped with serrated edges. The cheek teeth of herbivores vary. Some ornithischians have long crenulated crowns whilst other groups have broad grinding surfaces. In snakes and lizards the teeth are borne on the palatine and pterygoid bones as well as the jaws. *Sphenodon* has vomerine teeth but crocodiles have theirs restricted to the jaws.

ii Besides the venom glands of poisonous snakes and *Heloderma*, a variety of lingual, infralabial, palatine, supralabial and premaxillary glands can occur (Fig. 7.8).

iii Squamata have well-developed protrusible tongues but those of Chelonia and Crocodilia are restricted to the floor of the cavity. Snakes possess a cleft in the upper jaw which permits protrusion of the tongue even when the mouth is closed. The most elaborate tongues are those of chameleons.

(b) Alimentary canal

i Transport of food through the oesophagus is facilitated by its muscular and secretory nature. The region is considerably modified in egg-eating snakes, where it has a distensible anterior component, and in some Chelonia the oesophagus participates in crushing food.

ii The stomach is usually delimited into two sections, the corpus or fundus, and the pars pylorica. In most apodous genera the whole structure is elongated and in some snakes it is not differentiated from the oesophagus. Hydrochloric acid and pepsin are secreted into it and sand is sometimes swallowed by insectivorous genera to

act as a triturating agent. Such animals have gastric and intestinal chitinases.

iii The intestine is longest in turtles and shortest in snakes. The epithelium of the small intestine is simple in snakes and lizards but stratified in turtles and crocodiles. The intestinal juice itself contains few enzymes. Those secreted by the pancreas undergo adsorption on to a mucopolysaccharide at the intestinal surface. The large intestine may be subdivided into colon, cloaca and posterior cloacal canal.

iv The pancreas secretes amylases, lipases, chymotrypsin, trypsin, carboxypeptidase and an elastase.

RESPIRATORY SYSTEM

i Some chelonians use their pharynx as a supplementary site of gaseous exchange and soft-shelled turtles use their skin. The majority of reptiles rely upon pulmonary exchange. The presence of a hard palate in crocodiles and therapsids, the posterior situation of the internal nares and the long nasal passages, all presage the separation of the respiratory and alimentary tracts that occurs in mammals.

ii Most genera breathe by moving their ribs. In the Chelonia, where this is not possible, the mechanisms vary. In *Testudo* the major pumping action is provided by movements of the pectoral girdle pivoting on the dorsal and ventral articulations of the scapula and shell. *Chelydra*, the snapping turtle, can make similar movements but its principal pump involves those muscles that open and close the glottis, others that traverse under the viscera and yet others that act at the openings to the shell. When those below the viscera contract they compress the lungs and expel air. When those at the openings of the shell contract, the lungs expand.

URINOGENITAL SYSTEMS

(a) Excretory systems

i Adult reptilian kidneys differ from those of amphibians because the mesonephros has lost its excretory functions which are taken over by a new development of segmentally arranged tubules called, collectively, the metanephros. For a time the embryonic mesonephros does serve as a functional kidney but its place is then taken by an outgrowth from its hind end. Tubules grow out on each side, divide repeatedly, and give rise to

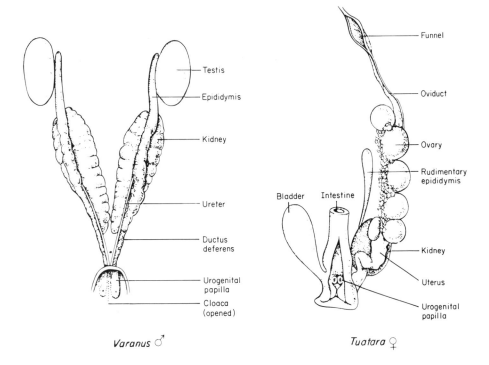

Varanus ♂   *Tuatara* ♀

Fig. 7.9 The male and female reproductive systems of reptiles.

the adult kidneys. Their ducts separate from the archinephric ducts and open independently into the cloaca alongside the archinephric ducts of the males or the oviducts of the females.

ii The kidney units also differ from those of amphibians in the smaller size of the glomeruli which reduces the total volume of urine that is produced and is an adaptation to water conservation. Indeed some snakes and lizards have aglomerular kidneys analogous to those of some teleosts. Terrestrial genera also substitute uric acid as the principal terminal product of nitrogen metabolism. In alligators and most aquatic genera ammonia remains the substance that is excreted but its toxic properties necessitate large volumes of fluid to be excreted as a diluent.

(b) The male system

Male reptiles retain the archinephric duct as the vas deferens. The mesonephros becomes the epididymis which is a mass of tubules that receive the vasa efferentia from the testis and act as a store for sperm before they

are delivered into the archinephric duct. The proctodaeum or posterior cloacal canal of all living reptiles apart from *Sphenodon* bears intromittent organs.

(c) The female system

i The female systems bear a close relationship to those of the foregoing classes. Traces of the archinephric duct persist as a strand of tissue, Gartner's duct, that terminates behind the ovary, and the mesonephros is represented by small, possibly functionless, masses of tissue known as the epoophoron and paroophoron.

ii The cleidoic egg is one of the prime adaptations to terrestrial life and fossil reptile eggs have long been known from Permian deposits. The egg differs from those of amphibians in the relatively large quantity of yolk which facilitates a prolonged embryonic life. During its passage down the oviduct it is first of all surrounded by albumen secreted by the proximal oviducal regions, then by shell membranes and finally by a shell. This is either tough and leathery or hard and calcareous, but it is always porous. The presence of a shell necessitates internal fertilization.

The innermost layer of albumen is denser than the outer layers, contains mucin fibres, and forms a thin covering over the egg outside the vitelline membrane. There is relatively little of this albumen in lizards but more is present in chelonians and crocodiles. The great mass of yolk modifies development and a blastoderm forms on one side of the egg with development proceeding as in birds.

ENDOCRINE GLANDS

(a) Hypophysis

An ovoidal neurohypophysis usually lies above the pars distalis but it can be asymmetrical. It is dorsolateral in some snakes and lateral in some fossorial genera. A solid structure in snakes, it is penetrated by the infundibular recess in other living forms. The adenohypophysis comprises an intermediate lobe that is closely associated with the foregoing and only joined to the pars distalis by a narrow cell bridge. In broad terms one can say that prolactin, ACTH, growth hormone, MSH, FSH, LH, and TSH secreting cells have all been identified.

(b) The thyroid gland

The reptilian thyroid is always encapsulated and connective tissue septa penetrate its interstices giving an argyrophilic fibre network around the follicles which vary in size from $50-300 \mu m$. Large parafollicular cells occur. The available data point to similar controls over metabolism and development as occur elsewhere. Thyroidectomy leads to a decrease in activity and finally death with indications of anaemia.

(c) The parathyroid glands

Unlike those of mammals the parathyroid glands of reptiles do not migrate and are found near to the thymus and ultimobranchial glands. Their structure resembles that in urodeles, birds and mammals. Essentially similar in all reptiles, they have a cord-like organization with sinusoids or capillaries between the cords. Follicular structures sometimes surround a lumen that is thought to represent a store of parathormone. Parathyroidectomy gives hyperexcitability and tetanic convulsions. Administration of exogenous parathyroid extracts to normal specimens gives hypercalcaemia, hypercalciuria and hyperphosphaturia.

(d) The thymus gland

There are two thymus structures variably located in the neck whose fine structure resembles that in mammals. Three cell types are noteworthy. Thymocytes measure from $4-6 \mu m$; epithelial cells have varying cytoplasmic characteristics, and hyoid cells of $20-30 \mu m$ diameter possess myofibrils around the nucleus. Many questions remain about the overall involvement in immune responses.

(e) The adrenal tissues

The various orders differ in the precise form and location of their inter-renal, adrenalin- and noradrenalin-producing tissues. However, the cells that are the equivalent of the mammalian medulla are all peripheral to the inter-renal tissue which is the homologue of the mammalian cortex. Many chelonians have the medullary tissue adpressed to the kidney but elsewhere it is near to the gonads. The relatively sparse studies on responses to exogenous corticosteroids or hypophysectomy suggest similar functions to those that are known in other taxa.

(f) The endocrine pancreas

In reptiles this lacks a sharp demarcation from the exocrine elements, usually lacks a capsule, and is associated with exocrine ducts. Two types of cells occur but their relationships to those of mammals remain unclear.

THE CARDIOVASCULAR SYSTEM

(a) The heart

i Most reptiles have a three chambered heart. The right atrium is always fully separated from the left one and both empty into the single ventricle. This is always subdivided by an incomplete septum but in crocodiles the septum is complete and the heart has four chambers. In this case the pulmonary and left systemic arteries leave from the *right* ventricle, the right systemic artery from the *left*. In the area of contact between the two systemic vessels there is a foramen, the foramen of Panizza, linking them.

ii Investigations in crocodiles suggest that the left systemic receives much of its blood from the right systemic via this foramen and not from the right ventricle. This reflects the relative pressure in the two systemics which is generally higher than that in both the right ventricle and the pulmonary artery so that the valve guarding the entrance from the right ventricle to the left systemic is prevented from opening. Oxygenated blood from the left atrium, therefore, enters the systemic circulation and deoxygenated blood from the right atrium goes to the lungs via the pulmonary artery.

(b) The arterial and venous systems

i In phylogenetic terms the internal carotids represent the third pair of arterial arches (Fig. 7.10), the systemics the fourth pair and the pulmonary arteries the sixth. A ductus caroticus linking the systemic and carotid arches is open in many reptiles but it is not a connection between the venous and arterial circulations.

ii The remainder of the circulation resembles that in amphibians. However, there is no renal portal and the posterior cardinal circulation is being taken over by the posterior vena cava. An anterior abdominal vein persists, short circuits some of the blood from the kidneys, and carries it from the pelvic region to the liver.

SENSE ORGANS

(a) Olfactory organs

The nasal cavity expands dorsally and the olfactory

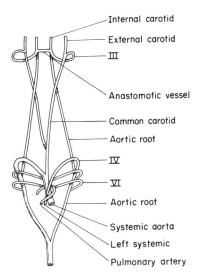

Fig. 7.10 The aortic arch derivatives of the grass snake. III, IV, VI aortic arches. (After Kerr.)

epithelium is restricted to its dorsal part. The ventral region of the cavity and a small anterior vestibule serve for the passage of air.

Jacobson's organ situated in the roof of the mouth and not far from the internal nares serves the general function of testing the chemical characteristics of food.

(b) Photoreceptors

i THE PINEAL ORGANS
In *Sphenodon* and many lizards the extracranial parapineal vesicle has an eyelike morphology. The cells of the roof form a thickened lens and photoreceptor cells predominate in the floor of the vesicle. The organ mediates thermoregulatory behaviour in lizards.

ii THE LATERAL EYES
The eyes vary in their development. Those of turtles and *Sphenodon* are the least elaborate, those of lizards the most elaborate and closely similar to those in birds. Those of crocodiles, despite their close affinity with the archosaurian ancestors of birds, are somewhat modified. A retractor bulbi muscle is always present, as is a nictitating membrane, and in both Chelonia and crocodiles a special muscle, the pyramidalis, is attached to a tendon that displaces the nictitating membrane. In snakes a transparent protective cover, or spectacle, is formed by fusion of the lids.

An overall view of the lizard's eye is provided by Fig. 7.11. Accommodation is achieved by changes in the shape of the lens. The ciliary body is long and its base lies within some 15 scleral ossicles which are situated close to the cornea. These ossicles are absent from snakes and crocodiles. Contraction of the ciliary muscle draws the

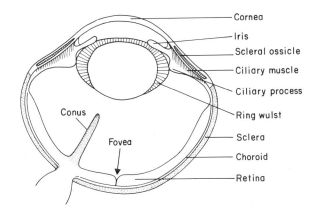

Fig. 7.11 Horizontal section through a lizard's eye.

ciliary body forward and its front part moves inward towards the axis of the eyeball thereby pressing on the annular fibrous pad of the soft lens. This deformation of the lens increases both its central axis and the curvature of the front and hind surfaces.

In the inner part of the vitreous cavity there is a highly vascular conus papillaris (c.f. the pecten of birds) which probably serves a nutritive function, nutrients diffusing to the retina, but it may also provide an enhanced ability to detect the position of moving prey as it breaks up the visual field.

(c) The ear

The position of the ear has already been mentioned in connection with the skull. The middle ear is still similar to that of amphibians. The inner ear includes sacculus, utriculus and semicircular canals. It still bears a lagenar macula but this is not the primary organ of hearing. The lagena is surrounded by a tubular extension of the perilymphatic system. Where this passes the papilla basilaris it is separated from the sensory cells by a basilar membrane. Vibrations from the oval window are transmitted by the perilymph and stimulate the sense cells via this membrane.

THE CENTRAL NERVOUS SYSTEM

(a) The medulla oblongata

The brainstem of reptiles, and the reticular formation in particular, is rather more differentiated than that of amphibians. The nuclei of the cranial nerves also vary in both their size and distribution in the different orders (Fig. 7.12). Vestibular and cochlear nuclei are well defined. Cochlear afferents pass to the magnocellular and angular nuclei within which they preserve a tonotopic organization. In the angular, the characteristic frequencies to which cells respond increase in the lateral to medial, anterior to posterior, and ventral to dorsal directions. Such differences are less clear in the magnocellular. In, for example, the caiman, responses occur to stimuli varying from 0.07 to 2.9 kHz.

(b) The cerebellum

Most adult reptiles have a unified cerebellum but that of crocodiles has anterior, median and posterior lobes which recall the early ontogenetic stages in birds and mammals. The overall appearance nevertheless varies. In some turtles it is a small flat plate whilst in others it is

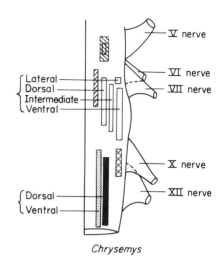

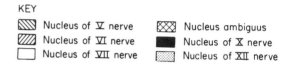

KEY

▨ Nucleus of V nerve		⊠ Nucleus ambiguus	
▨ Nucleus of VI nerve		■ Nucleus of X nerve	
☐ Nucleus of VII nerve		▦ Nucleus of XII nerve	

Fig. 7.12 Medullary nuclei of the cranial nerves in the chelonian genus *Chrysemys*.

huge; many lizards and snakes have a forward tilted structure (Fig. 7.13) whose molecular layer lies anteriorly and whose granular layer lies posteriorly. Early experimental investigations demonstrated that differing postural patterns followed stimulation of different regions.

(c) The mesencephalon

The overall organization of the midbrain is closely comparable with that in other orders but the optic tecta of many lizards and snakes, together with those of crocodiles, are elaborate and comprise some fourteen distinct layers. In lizards these layers are best represented in diurnal genera and are very much reduced in fossorial ones. Fig. 7.14 shows that a clear retinotopic organization results in different retinal areas being represented at particular foci.

(d) The diencephalon

The hypothalamus and epithalamus are similar to those in foregoing classes. The thalamus is complex. The rotund nucleus represents a relay station on an ascend-

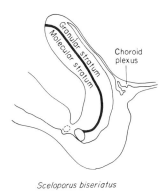

Sceloporus biseriatus

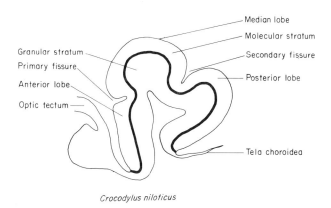

Median lobe
Molecular stratum
Secondary fissure
Posterior lobe
Granular stratum
Primary fissure
Anterior lobe
Optic tectum
Tela choroidea

Crocodylus niloticus

Fig. 7.13 The cerebella of a lizard and the nile crocodile.

ing tecto-rotundo-telencephalic visual pathway. The lateral geniculates lie within a more direct retino-geniculate-telencephalic pathway. The precise function of other nuclear moieties remains to be determined.

(e) The hemispheres

A stylized transverse section of one hemisphere of a reptile is presented in Fig. 7.15. There has been much discussion about the homologies with mammalian structures. It is clear that pyriform, general and hippocampal cortex covers the dorsal and dorsolateral regions and, by analogy with the situation in birds, it is probable that the hyperstriatal and neostriatal moieties represent solid developments from the same population of cells that gives general cortex and, in mammals, neocortex. Only the paleostriatal structures are in any way homologous with mammalian basal ganglia. Most of the hemisphere regions receive input from olfactory sources but both photic and somatosensory effects can be detected in the general cortex.

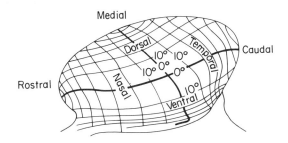

Fig. 7.14 Diagram showing the retino-tectal projections of the alligator. (After Heric & Kruger.)

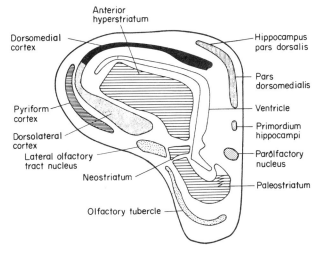

Anterior hyperstriatum
Dorsomedial cortex
Hippocampus pars dorsalis
Pars dorsomedialis
Ventricle
Pyriform cortex
Primordium hippocampi
Dorsolateral cortex
Parolfactory nucleus
Lateral olfactory tract nucleus
Neostriatum
Paleostriatum
Olfactory tubercle

Fig. 7.15 Diagrammatic transverse section of a reptile hemisphere.

ADAPTIVE RADIATION

(a) Anapsida

The group of genera that are attributed to the Cotylosauria, or stem reptiles, has been wittled down over the years. The Seymouriamorpha are now clearly seen to be amphibians and the Diadectomorpha have also had an amphibian relationship ascribed to them. In broad terms this leaves the Captorhinomorpha and Chelonia as definitive, if rather unrelated, anapsids. The captorhinomorphs are primitive and rather early fossil forms. The Chelonia are known in time from the Triassic to the present and comprise a coherent assemblage that is divisible into the sub-orders Pleurodira and Cryptodira respectively by virtue of either a lateral or a vertical bending of the neck when the head is withdrawn into the shell. The cryptodires form the most

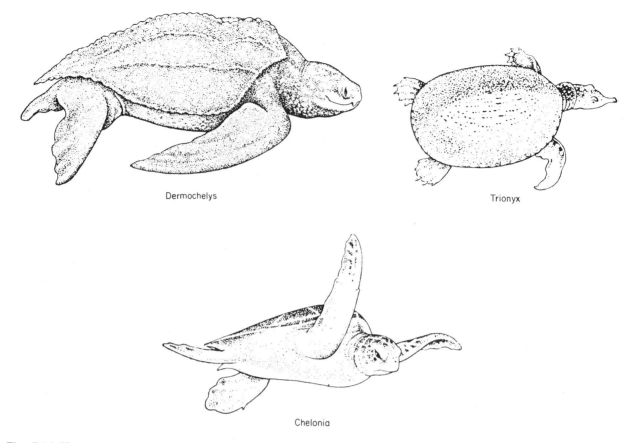

Dermochelys

Trionyx

Chelonia

Fig. 7.16 Three genera of chelonian.

diverse component and include some four superfamilies (Fig. 7.16).

(b) Synapsida

i The synapsid stock from which mammals evolved achieved dominance during the Permian, and then later sank into relative obscurity in the face of diapsid and parapsid diversification. The pelycosaurs (Fig. 7.17), of which *Dimetrodon* is the best known genus, often had prominent sails on their dorsal surface. These may have served as a primitive heat control mechanism as a result of arteriolar constriction and dilation, but may also have served supplementary functions such as the directional localization of sound.

ii During the Permian various phyletic lines exhibited a progressive increase in the contribution made by the dentaries to the lower jaw (Fig. 7.18). Additional units,

such as the articular, angular and surangular bones, became relegated to a posterior and relatively insignificant position. These tendencies culminated during the Triassic in forms like the tritylodonts and ictidosaurs whose general form parallelled that of mammals such as the multituberculates and rodents. Various synapsid fossils have impressions in the mouth region. As these compare closely with those underlying vibrissae in some modern mammals they suggest the presence of hair. Such genera also have restricted rib cages which suggests the presence of a diaphragm. They lie on the reptile/mammal boundary and were probably homoiothermal but perhaps not viviparous.

(c) Ichthyopterygia

The fishlike appearance of ichthyosaurs emphasizes their extreme specialization to an aquatic way of life. They show no close similarities to other groups of

Fig. 7.17 A reconstruction of a pelycosaur.

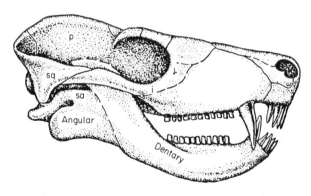

Fig. 7.18 The skull of *Bauria*, a theriodont reptile. (After Broom.) Abbreviations: 1, lachrymal; sa, surangular; p, parietal; sq, squamosal.

reptiles and the presence of what appear to be young in the pelvic region of various fossils suggests that they may have been viviparous. The taxonomic divisions within the subclass reflect the differences between broad or narrow fins—the latipinnates or longipinnates. In those

cases where an outline of the body is preserved it is clear that they had vertical tail flukes, not horizontal ones as in whales, together with dorsal flukes comparable with those of dolphins.

(d) Lepidosaurs

i   Most classifications distinguish between two diapsid groupings, the lepidosaurs on the one hand and the archosaurs on the other. Three groups are habitually placed in the Lepidosauria—the ancestral Eosuchia, the Rhynchocephalia and the Squamata. Only rather minor characteristics separate primitive rhynchocephalians from eosuchians such as *Youngina*. This last had two temporal vacuities, a low lateral one and a large upper one (Fig. 7.19). *Sphenodon*, although resembling a lizard, retains primitive features such as the prominent pineal eye, and the presence of two definitive temporal vacuities forms a sharp demarcation from lizards where the lower temporal bar is lost.

ii   The Squamata includes both lizards and snakes. The former are known from the Jurassic and include the

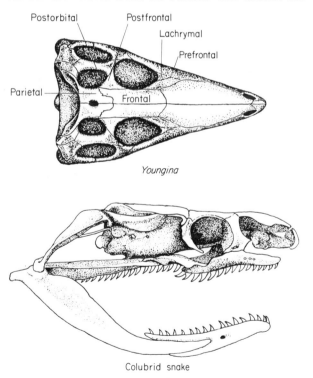

Fig. 7.19 Dorsal view of the skull of *Youngina*, and lateral view of the skull of a colubrid snake.

aquatic mosasaurs of the Cretaceous. Their adaptive radiation is represented in Fig. 7.20. All the really powerful snakes are either boas or pythons but not all of these are large in size. Some retain vestiges of a pelvic girdle. There are, in addition, several other families of non-poisonous snakes and several families lack fangs. Such categories include the blind, worm and shield-tailed snakes. The rear-fanged genera, opisthoglyphs, are also variously related to genera with and without front fangs. Specialized teeth at the back of the jaws inject venom during a chewing action. Definitive, tubular, front fangs occur in cobras, coral snakes, vipers, pit vipers, etc.

(e) Archosaurs

i THECODONTS
The sub-class Archosauria includes the primitive theco-donts together with the crocodiles, dinosaurs and pterosaurs. The thecodonts themselves include a varied array of animals—Proterosuchia, Pseudosuchia, phyto-saurs and aetosaurs. The pseudosuchians were small carnivorous reptiles with a superficial similarity to lizards but having their sharp teeth set in sockets. Some were fully bipedal and all combined generalized archosaurian features with various specializations.

ii SAURISCHIA
The animals customarily referred to as dinosaurs are of two types that can be differentiated on the basis of their pelvic structure. There has been some considerable discussion about the possible existence of homeothermy and endothermy in such forms. The saurischians include bipedal carnivorous genera, grouped together as theropods, and secondarily quadrupedal herbivorous ones—the sauropodomorphs. All have a triradiate pelvic structure. The theropods include the lightly built coelurosaurs, their derivative genera such as *Struthio-mimus* (Fig. 7.21), and the large carnosaurs such as *Tyrannosaurus*.

The herbivorous sauropodomorphs include most of the well-known genera of herbivorous dinosaurs. *Plateosaurus* is an early pro-sauropod genus of which several entire skeletons are known from Upper Triassic deposits in Germany, France and South Africa. The sauropods themselves include some of the largest ter-

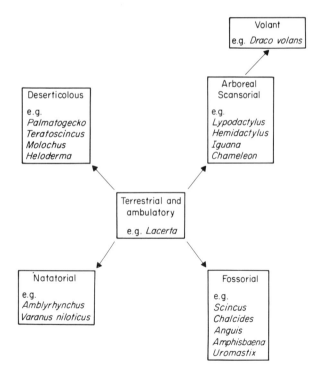

Fig. 7.20 The adaptive radiation exhibited by modern lizards.

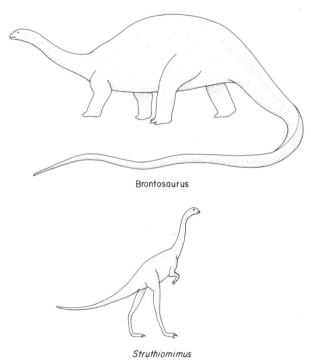

Fig. 7.21 Reconstructions of two well-known saurischians.

restrial animals that have ever existed and there has been much discussion about the degree to which they had amphibious habits and relied, in part, on water to support their bulk. They include *Apatosaurus* (Brontosaurus) and *Diplodocus* (Fig. 7.21).

iii ORNITHISCHIA

The quadrupedal and herbivorous ornithischians differ from the saurischians in their quadriradiate pelvis (cf. page 95). Very varied in form, they include the generalized ornithopods such as *Iguanodon*, the plated stegosaurs, the armoured ankylosaurs, the horned ceratopsians and other groups such as the hadrosaurs (cf. Fig. 7.22).

iv CROCODILIA

The order Crocodilia consists of the following families: Crocodylidae with the genera *Crocodylus*, *Osteolaemus* and *Tomistoma*; Alligatoridae with the genera *Alligator*, *Caiman*, *Melanosuchus* and *Paleosuchus*, and Gavialidae with the genus *Gavialus*. In crocodiles the teeth of the upper and lower jaws are more or less in line and, therefore, all visible when the jaws are closed. The fourth tooth of the lower jaw, which is considerably longer than its neighbours, fits into a constriction of the snout and can be seen pointing upwards almost like a tusk. Alligators and caimans have the lower teeth inside the upper row and the enlarged fourth tooth fits into a pit in the upper jaw so it is not visible.

The various genera and species vary considerably in both appearance and adult size. In broad terms the young feed on insects and graduate first to small fish and then to larger prey as they increase in size. The two smooth-fronted caimans, *Paleosuchus palpebrosus* and *P. trigonatus* are both very small and only measure 90–150 cm in length. In contrast, specimens of the Orinoco crocodile, *Crocodylus intermedius*, can measure 6–8 metres, the black caiman, *Melanosuchus niger*, can attain 6 metres, and various species average around 4 metres. Some large individuals, often living in estuaries, have been recorded at over 8 metres.

Their body temperature fluctuates around 25°C ± 3°C. When too hot they open their mouths and gape widely and the loss of water occurring under these circumstances has been cited as up to 20 % of the body weight. At noon they may crawl under bushes or return to the river. Birds like the spurwing plover, *Haplopterus spinosus*, remove leeches, etc., from the jaws of Nile crocodiles, and similar mutualistic relationships occur between the American crocodile and a species of fish.

Females select a nesting site in which to deposit their eggs and lay them at the bottom of a pit which can be up to 0.5 metres deep. They then remain in the vicinity and guard the site because the eggs are preyed upon by monitor lizards.

(f) Parapsida

i Finally there are the forms with a single superior temporal vacuity comparable with the upper one in

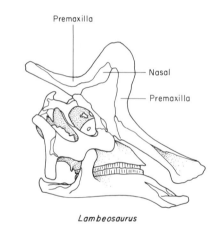

Lambeosaurus

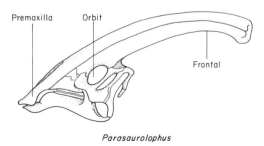

Parasaurolophus

Fig. 7.22 The skulls of *Lambeosaurus* and *Parasaurolophus*.

Fig. 7.23 Reconstruction of a plesiosaur.

diapsids. These include small lizard-like forms such as the Permian *Araeoscelis* or the long-necked *Tanystropheus*.

ii   The plesiosaurs evolved from some such ancestry. They, together with nothosaurs and placodont reptiles, are classed as Sauropterygia. This is probably a somewhat loose grouping. The limbs of nothosaurs and placodonts were still suitable for walking on land but those of the plesiosaurs were long paddle-like structures.

FURTHER READING

BELLAIRS A. (1969) *The life of reptiles*, 2 vols. Weidenfeld and Nicholson Ltd., London.

GANS C. (ed.) (1969) *The biology of reptiles*, 10 vols. Academic Press, London.

GOIN C.J. & GOIN O.B. (1971) *Introduction to herpetology*. W.H. Freeman & Co., New York.

PORTER K.R. (1972) *Herpetology*. W.B. Saunders and Co., Philadelphia.

ROMER A.S. (1956) *Osteology of reptiles*. University of Chicago Press, Chicago.

8 · Class Aves

THE BIRDS

Synopsis

Birds are usually ascribed to some 30 rather independent orders, many of which are surveyed on the following pages 122–130. However, for convenience it is possible to group these orders in the following way.

Class Aves
  Subclass Archaeornithes—*Archaeopteryx*
  Subclass Neornithes (Modern birds)
    Superorder Paleognathae—Birds with a primitive palate such as ostriches, rheas, emus and kiwis.
    Superorder Neognathae—Modern flying birds.

N.B. The penguins are sometimes removed to a separate superorder, the Impennae, and the order Tinamiformes, the tinamous, combines a primitive palate with the carinate structure of the sternum seen in neognaths.

INTRODUCTION

Birds, like mammals, are endothermic, producing heat from metabolism, although they maintain their body at the slightly higher temperature of c. 39°C (102°F). However, whilst mammals are derivatives of the synapsid stock, the close similarities which exist between birds and diapsid reptiles are universally acknowledged to reflect a common ancestry. Birds, like crocodiles, are descended from the archosaurian stock. The earliest known representative is the fossil genus *Archaeopteryx* which was discovered in the Jurassic deposits of the Solenhofen area of Bavaria in 1862. The impressions of feathers clearly established it as a bird, but the presence of sharp pointed teeth and a tail of some 20 vertebrae pointed to its reptilian ancestry.

INTEGUMENT

i Avian skin is much thinner and more delicate than that of mammals. It is attached to the musculature in relatively few places but has extensive attachments to the skeleton. The epidermis comprises a layer of living and a layer of cornified cells whilst the dermis is thin and rather uniformly fibrous. Scales cover the distal part of the hind limbs in many species.

ii The beak, bill or rhamphotheca is a keratinized epidermal structure which is continuously replaced. The upper and lower parts are generally equal but the upper portion is the longer in birds of prey. Several distinct horny plates occur in the albatross or fulmar but in the majority of birds these are fused to form one undivided sheath. Although teeth are absent from all modern birds, a calcareous protuberance develops on the upper surface of the bill during the embryo stage. This is known as the 'egg-tooth' and is used to break the egg shell. The bills of some adult birds, such as the fish-eating goosanders and mergansers, bear serrations which help to grip the prey. In those ducks that feed in soft mud the serrations take the form of horny lamellae which, together with the thick fleshy tongue, provide an efficient sifting apparatus. This attains its maximum development in the shovellers. Many variations in bill form occur in the various orders.

iii Feathers form the body covering. There are six main types—contour, down, filoplume, bristle, powder-down and semi-plume—and they are replaced, by moulting, at least once a year. Contour feathers are embedded in a germinative follicle by a short tube, the calamus, into the base of which a dermal papilla protrudes. The main shaft or rachis bears two series of stiff filaments, the barbs, at approximately 45°. These barbs bear barbules, again at 45° (Fig. 8.1), and the barbules of adjacent barbs cross each other at 90°. The hooks on the barbules engage to form the vanes.

The barbs of down feathers are devoid of hooked barbules, and bristles merely comprise a stiff rachis. Powder-down feathers shed minute particles of keratin and are of particular significance in herons where they are associated with the presence of a special pectinate digit on each foot which is used for preening. Semi-plumes have a fluffy vane. Various more specialized feathers occur in different species. Owls, for example, have emarginate feathers that facilitate a silent flow of air over the wings.

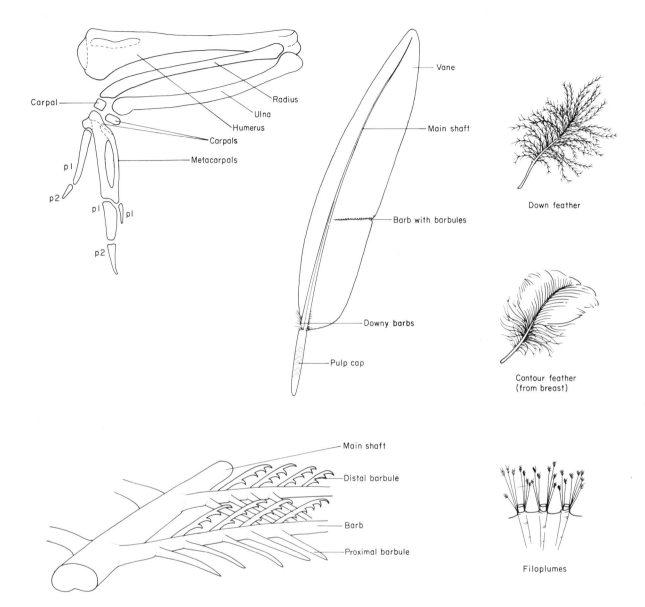

Fig. 8.1 The bones of a bird's wing together with feather types and a diagram showing the interrelationships of barbs and barbules.

SKELETAL SYSTEM

(a) The skull (Fig. 8.2)

i The skull bears a single occipital condyle and the sutures between individual bones are obliterated by ankylosis at an early stage. The roof of the cranial cavity is largely formed by frontal, parietal and supraoccipital elements; the floor by basioccipital, basisphenoid and parasphenoid ones. Squamosal, orbitosphenoid and prootic bones contribute to the lateral walls and the facial region is mainly made up of premaxillary and

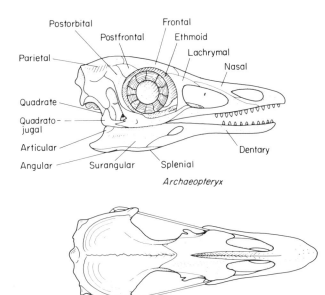

Fig. 8.2 Side view of the skull of *Archaeopteryx* and a dorsal view of the skull of a goose.

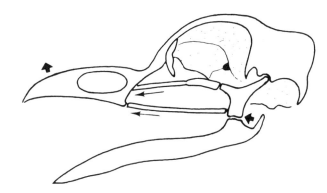

Fig. 8.3 A bird's skull showing the kinesis which results from depression of the lower jaw. (After King & McLelland.)

nasal bones. The large size of the eyeballs is associated with a thin mesethmoid which gives the interorbital septum. The lower jaw primitively comprises six bones.

ii A craniofacial hinge, between the facials on the one hand and the premaxillaries and nasals on the other, can permit an increase in the size of the gape in some species. The ventral processes of the quadrate bones are connected to the premaxillae by way of a thin jugal arch laterally, and by the palatine and pterygoid bones medially. The quadrates rotate in response to the forces generated by ventral displacement of the lower jaw and displace the facial region of the skull to give *kinesis*. This is pronounced in many parrots (Fig. 8.3).

(b) The axial skeleton

i Most vertebral, sternal, pelvic and costal bones contain diverticula of the airsacs—are pneumatized. There are very variable numbers of vertebrae, with 39 in some small passerines and 63 in the swan. Of these 7–25 can contribute to the cervical region. The thoracic region comprises a fused structure, representing 2–5 vertebrae, together with 2 or 3 free vertebrae. The synsacral region of *Archaeopteryx* was composed of 4 vertebrae but represents from 10–23 in extant forms.

Complementarily the caudal region comprised 22 vertebrae in *Archaeopteryx* but is represented by a pygostyle composed of about 6 fused vertebrae in modern genera.

ii A series of paired ribs form the lateral wall of the thoracic cage. Each is composed of a dorsal vertebral derivative and a ventral sternal one. Their number varies. Three pairs occur in the Bucerotidae; eight pairs occur in *Cygnus*. The form of the sternum divides birds into the 'Ratites' which have a platelike structure, and the carinates in which it bears a prominent ventral keel or carina. These two groups broadly correspond to the paleognathous genera, apart from the tinamous, on the one hand, and all other birds on the other.

iii The pectoral girdle has long, laminar scapulae that are attached to the ribs by muscles and ligaments. The clavicles are united to form the well-known furcula or wishbone. In the pelvic girdle the lack of any articulation of the pelvic bones in the midline permits the passage of the large-shelled eggs.

(c) The appendicular skeleton

i An avian wing is represented in Fig. 8.1. The humerus rotates away from the body during flight but lies against the thoracic cage when at rest. It receives the insertions of the supracoracoideus muscle which produces the upstroke, and the pectoralis muscles which produce the downstroke. The proximal region is penetrated by airsacs. Flexion and extension of the elbow joint are limited to the plane of the wing surface. The ulna is more massive than the radius and some flight feathers are attached to it by connective tissue.

ii The manus is reduced. Two bones represent proximal carpals and the distal carpals, free in the young, are later fused into a carpometacarpal unit. The digits are reduced to II, III and IV which each bear claws in the ostrich. One or two claws occur in the young of some nidifuges, such as the hoatzin or rails, but they are generally absent.

iii Besides the acetabular ball and socket joint, the trochanter of the hindlimb articulates with an antitrochanter on the ilium. This assists balancing when the bird is standing on one leg. A unified tibiotarsus is formed from the tibia and a proximal row of tarsal bones whilst the fibula is a slender rod fused to the tibiotarsus. A tarsometatarsus is formed from the distal row of tarsals and the three main metatarsals. There are four digits with two, three, four and five phalanges in many ground-living birds. The common pattern is three front toes and a backwardly directed hind toe. In passerines this hind toe is strong and well-developed for perching. Furthermore, an automatic muscular mechanism causes all four toes to close tightly round the perch. Skylarks are unusual amongst passerines as they have an exceptionally long claw on this hind toe that helps them to balance when running.

Climbing birds, such as woodpeckers, have two front and two hind toes. This zygodactylous condition also occurs in cuckoos. Birds such as swallows and swifts, which spend much time on the wing, have small toes, those of swifts all pointing forwards. Many wading birds have webbed feet and in those of pelecaniform genera the four toes are again forwardly directed.

THE MUSCULATURE

i Closure of the jaw involves pseudotemporalis muscles. Muscles of the hyobranchial region protract and retract both the tongue and larynx, whilst sternohyoideus elements assist the sternolaryngeal muscles in oscillating the syrinx during vocalization. The trunk musculature is very much reduced in the thoracic and synsacral regions. Gaseous exchange is accomplished by inspiration using external intercostals and the triangularis sterni. Expiration involves the internal intercostals together with abdominal muscles.

ii The rhomboideus and latissimus dorsi muscles protract and retract the wing. Extension of the shoulder joint is accomplished by the deltoideus, extension and flexion of the elbows by the triceps and biceps brachii. The fine control of the wingtip which is so important during flight is under the control of interosseous muscles together with both the adductors and abductors of the digits.

iii The protractors and retractors of the legs act by flexing or extending the hip joints. The gastrocnemius extends the intertarsal joint which is flexed by, for example, the tibialis cranialis. Extensor digitorum longus and other related muscles extend the digits.

FLIGHT

i The wing is an aerofoil and generates lift. In level flight at a steady speed the net result of flapping the wings is a force acting vertically upwards and balancing the weight of body. The relative airflow (V_r) at a point halfway along the wing is the resultant of that due to flapping motion (V_f) and that due to the birds forward speed (V). The relative wind is inclined upwards so that the net aerodynamic force (R) acting on it, and the resultant of lift and drag, can be inclined forwards of the vertical and provide a temporary thrust. At more distal points on the wing the vertical component of the relative airflow is larger, giving a larger component of forward thrust (Fig. 8.4).

ii Forward flight in a gliding bird is effected by fore and aft movements of the wings which can be rotated at

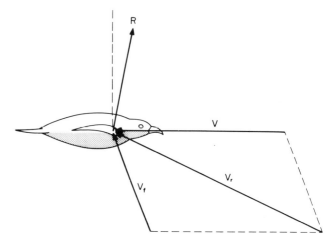

Fig. 8.4 Air flow and net aerodynamic force (R) in the middle of a downstroke giving a propulsive thrust. (From Pennycuick.)

the shoulder, elbow and wrist. To trim nose-down wings are shifted back and to trim nose-up they are moved forwards. The tail is used to supplement control movements of the wings, especially during rapid manoeuvres. It is spread and depressed at low speeds, especially during landing and take off, when it helps to suck air downwards. In swifts, swallows, terns and frigate birds the forked tail enables slow flight and hovering. The feet provide drag with limited lift and the most effective airbrakes comprise the webbed feet of aquatic species.

Many birds soar on a rising air current either by cliffs or on thermal currents evoked by solar heating of the air.

THE DIGESTIVE SYSTEM

(a) The oropharynx

There is no soft palate, so the pharyngeal region lacks any division into oral and nasal components. The choanal opening is a slit in the hard palate. The tongue, supported by the hyobranchial apparatus, is long and protrusible in humming-birds (Trochiliformes) or woodpeckers (Piciformes), relatively thicker in game birds (Galliformes). In the woodpeckers its muscles originate on a pair of curved rods which start on either side of the windpipe, continue upwards over the top of the head, where they meet, and then continue forwards in a deep groove in the skull and turn to one side before entering the nasal cavity. The muscles attached to these rods enable them to be pulled backwards, downwards and forwards, and thus the tongue is thrust in and out very rapidly. Sap-sucking woodpeckers have a shorter tongue with a brushlike tip.

Birds have relatively few tastebuds and salivary glands are most prominent in species with a dry diet, such as graminivores, or in woodpeckers where the secretion assists in capturing insects.

(b) The stomach

The avian stomach usually has two clear divisions, a glandular, proventricular component and a muscular gizzard (Fig. 8.5). The main ducts of the glands are lined by mucous cells and glandular alveoli have only one type of cell whose fine structure is similar to that of both the parietal and peptic cells of mammals. They are designated oxyntopeptic cells. The proventriculus is distensible in albatrosses, gulls, etc., and gastric proteolysis of graminivores takes place in the gizzard.

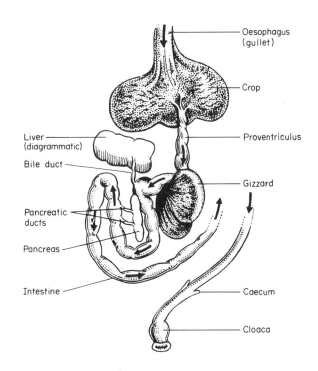

Fig. 8.5 The alimentary canal of a bird.

(c) The intestines

i The duodenum is a narrow U-shaped loop and it is followed by the jejunum and ileum arranged in similar loops. The vitelline diverticulum is a short, blind remnant of the yolk sac.

ii The large intestine comprises paired caeca and a short, straight rectal component. Caeca arise in most birds at the junction between the ileum and rectum and this last opens into the coprodeal region of the cloaca by way of a sphincter muscle.

THE RESPIRATORY SYSTEM

i The nasal cavity, which is involved in both heat and water conservation, contains scrolled or laminate conchae. The middle one of these is the largest and is branched in emus. The expired air is cooled down when passing through the region so that water condenses out and is retained.

ii The lungs are situated dorsally and have a total volume which is only about one tenth of that in mammals of equivalent weight. Primary bronchi divide to form secondary ones and, a few millimetres from their origin, these give rise to parabronchi. These are tubes of uniform diameter. Six pairs of airsacs occur in the embryo but in the majority of birds two of these pairs fuse around the time of hatching. As a result, the adult has a median cervical sac; a clavicular sac derived from four primordial ones; a pair of anterior thoracic, a pair of posterior thoracic and a pair of abdominal sacs (Fig. 8.6).

iii The direction of airflow is constant. It passes from the mediodorsal secondary bronchi, through the para-bronchi and into the medioventral secondary bronchi. Caudal airsacs receive relatively fresh air during each inspiration at a time when the more anterior sacs receive air which has already passed over the gaseous exchange tissues. During expiration air is expelled from the anterior airsacs into the primary bronchus and thence to the trachea. At the same time the posterior sacs expel their air into the lungs where it traverses the gas exchange tissues.

iv Birds have a larynx but this lacks vocal chords. Song is a product of the syrinx. This is usually situated at the point where the trachea bifurcates but, less commonly, it is in the lower part of the trachea or in the bronchi. In the most common form the lower rings of the trachea coalesce giving a resonating chamber within which there occur one or more tensely stretched membranes over which air passes. The tension is controlled by muscles. There are, however, many departures from this typical form. Cassowaries lack a syrinx yet are capable of producing many sounds. In the emu the tracheal rings do not unite but leave an aperture through which the

lining membrane emerges to form a resonating sac. The velvet scoter has two chambers. In many species the windpipe itself is extraordinarily long and contributes to sound production. It is coiled in the breast muscles of the semi-palmated goose, some sandpipers and some passerines; in a cavity of the carina in swans and some cranes, or beneath the lungs. Such long tracheae are usually the prerogative of the male but can occur in both sexes and, very rarely, in the female alone.

THE URINOGENITAL SYSTEMS

(a) The kidney

i The avian kidney (Fig. 8.7) comprises numerous lobules which are enclosed by their collecting tubules. These are, therefore, interlobular, whereas the efferent vein draining the lobule is intralobular. There are two types of nephron. The predominant or cortical type lacks a loop of Henle, is confined to the cortical region and recalls the reptilian condition. About half of its length is formed by a proximal convoluted tubule lying in an N-shaped configuration. The medullary type has a loop of Henle which penetrates the medulla. Juxtaglomerular structures include a *macula densa* or thickening of the epithelium in the distal convoluted tubule at the point of contact with the afferent arteriole.

ii The ureter originates within the anterior division of the kidney, passes in a groove on the ventral surface of the middle and posterior divisions, and opens into the

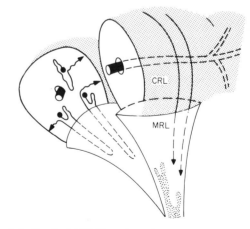

Fig. 8.7 Cortical (CRL) and medullary (MRL) regions of renal lobules. Three nephrons are shown on the left; the upper is a cortical one and the lower two are medullary nephrons with medullary loops. (Modified from Johnson.)

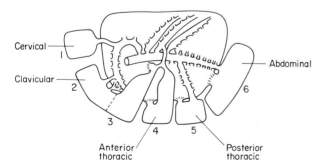

Fig. 8.6 The right lung of a fowl showing the six embryonic airsacs and the final adult condition. The second and third embryonic pairs are fused to each other and across the midline. (After King & McLelland.)

urodeum. The 'bladder' of the ostrich is probably a capacious diverticulum of the cloacal bursa.

iii Birds, like mammals, excrete urine which is hypertonic to the blood. However, they resemble reptiles in being uricotelic. This excretion of uric acid may be particularly related to overcoming the difficulties associated with a prolonged development in a cleidoic egg.

(b) The male reproductive system

i In birds it is the male which is the homogametic sex. Furthermore, during development the left testis acquires many germ cells from the right primordium and is, therefore, the larger organ in the mature animal. The actual dimensions vary greatly with the sexual cycle, as does the colour. A resting testis may be black due to the presence of melanoblasts. The surface is covered with a tunica albuginea, anastomoses between the individual seminiferous tubules are more common than in mammals, and interstitial cells between the tubules are rather sparse. It is these that are the source of androgens.

ii The epididymis is much less conspicuous than in mammals. The numerous coiled efferent ductules open into a straight epididymal duct, the vas deferens forms a tight zigzag near to the midline and terminates in a spindle-shaped receptacle at the urodeum.

iii There are no accessory glands comparable with the prostate, seminal vesicles or bulbo-urethral glands of mammals. The phallus is derived from the urodeal wall, the proctodeal floor and the ventral lip of the vent. It is fully protrusible in ratites and Anseriformes.

(c) The female reproductive system

i It is usually the left organs that are the functional structures in the adult. In young birds the ovary is clearly delimited into an outer cortex, which contains the oocytes, and an inner medulla, but this distinction is obscured later. Avian primary oocytes include the largest individual cell in the world. In all species the primary oocytes are suspended by a vascularized and muscular stalk, and enclosed by the wall of the follicles which can be differentiated into six layers:

    an inner layer;
    the stratum granulosum;
    the theca interna;
    the theca externa;
    connective tissue;
    epithelium.

ii The left oviduct has five parts, viz., the infundibulum, magnum, isthmus, shell gland and vagina. Newly released secondary oocytes pass into the infundibulum and the second maturation division occurs in the oviduct. The chalaziferous layer of albumen is deposited whilst the egg is in the infundibulum, the bulk of the albumen whilst it is in the magnum. Inner and outer shell membranes are laid down in the isthmus. Aqueous solutions are added (plumping) and then calcification occurs.

iii The ovary and its derivatives are involved in various endocrine functions. Oestrogens are secreted by cells of the follicular wall. Androgens are secreted by interstitial cells, and progestogens by the follicular components after rupture. In both sexes the sequential peaks in the levels of circulating hormones are intimately related to the different phases of courtship and nest-building behaviour. They exert these controls by influencing the activity of preoptic cells in the brain (q.v.).

OTHER ENDOCRINE GLANDS

(a) The hypophysis (Fig. 8.8)

The adenohypophysis includes a pars tuberalis and a pars distalis. It is the latter which is the larger component and its cells are arranged in a follicular manner. The differential occurrence of varying cell types permits a differentiation of rostral and caudal zones. These cell types have particular hormonal secretions:

    FSH is secreted from Type I cells;
    TSH from Type II cells;
    LH from Type III cells;

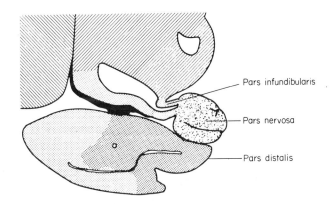

Fig. 8.8 Overall distribution of the hypophysial components of the partridge.

prolactin from Type IV cells;
STH from Type V cells.

It is worth noting that prolactin has very diverse functions, and circadian rhythms in its concentration underlie, for example, the early morning and late afternoon feeding periods.

The neurohypophysis comprises a median eminence, a pars infundibularis and a pars nervosa. Vasotocin and oxytocin produced in the hypothalamic nuclei are transported to the neurohypophysis via the hypothalamic tract and then stored there prior to release.

(b) The thyroid glands

These are follicular structures that are derived from the pharynx at the level of the first and second pharyngeal pouches. Each follicle contains thyroglobulin in its lumen. Both thyroxin and triiodothyronine stimulate metabolic processes and thereby regulate heat production and body growth as well as controlling moulting.

(c) The parathyroid glands

These originate from the third and fourth pharyngeal pouches and comprise anastomosing cords of columnar chief cells which secrete parathyroid hormone. This mobilizes bone calcium and raises blood calcium levels.

(d) The ultimobranchial glands

Arising from the sixth pharyngeal pouch, these glands lie close to the parathyroids. They consist of rather scattered eosinophilic, or C, cells that secrete calcitonin which inhibits the mobilization of bone calcium.

(e) The adrenal glands

Once again these lack the clear differentiation into cortical and medullary components that one sees in mammals. Cells which originate respectively from the neural crest and from mesoderm are intermingled. The former secrete noradrenalin and adrenalin; the latter secrete corticosteroids which control electrolyte balance and carbohydrate metabolism.

(f) Islets of Langerhans

There are two types of islets. Alpha islets are composed of α_1 and α_2 cells, the latter secreting glucagon which raises blood glucose levels. Beta islets are composed of β and α_1 cells. β cells secrete insulin which speeds up the phosphorylation of glucose, facilitates its transport across cell-membranes and into cells, and thereby lowers blood glucose levels.

THE CARDIOVASCULAR SYSTEM AND RELATED STRUCTURES

(a) Arterial components

i The aorta is the right, fourth arterial arch and the right dorsal aorta of the embryo (cf. mammals). The first branches of the aorta are the right and left coronary arteries, the right and left brachiocephalic trunks then follow. Each of these divides to give a common carotid and a subclavian artery which supplies the wing and breast musculature. A derivative brachial artery supplies the proximal parts of the wing whilst radial and ulnar arteries supply more distal regions. The short common carotid gives rise to a vertebral artery and a long internal carotid.

ii The descending aorta supplies paired intercostal and synsacral arteries to the trunk; the coeliac artery to the stomach, small intestine, caeca and pancreas; the anterior mesenteric artery again to the intestinal regions, and anterior, medial and posterior renal arteries to the kidneys (Fig. 8.9). Testicular and ovarian arteries

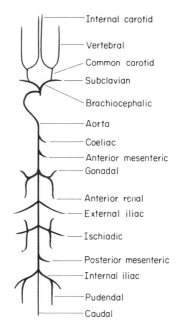

Fig. 8.9 Diagram of an avian arterial system.

usually arise from the anterior renal vessel. The hind legs are supplied by external iliac vessels, which continue distally as the femorals, and by the larger ischiadics which continue distally as the popliteal and anterior tibial arteries.

Internal iliac arteries supply the tail region, by way of lateral caudals, and the vaginal and cloacal regions via the pudendal. In the midline the hind extremity of the aorta gives the median caudal artery.

(b) Venous components

i The posterior vena cava receives the hepatic veins from each side (Fig. 8.10). Further back it has received the ovarian and testicular derivatives and the large common iliac veins. The hind legs are drained by large external iliac veins and small ischiadic veins. The first-named is an anterior prolongation of the more distal popliteal and tibial veins which fuse giving the femoral component and then the iliac.

Two hepatic portal veins drain the stomach and intestinal regions, although the principal drainage of the large intestine is provided by the anterior mesenteric vessel.

ii Anteriorly the subclavian and jugular veins of each side fuse to give the anterior venae cavae. The intra-cranial region lacks precisely defined veins and is drained via large venous sinuses. Blood within the sagittal, petrosal, transverse and annular basilar sinuses within the cranial cavity converges onto the occipital region where it leaves the cavity and joins the jugular by way of occipital veins.

(c) The lymphatic system

i Lymphatic vessels are less numerous than in mammals. There are typically two to each blood vessel. Retrograde flow is prevented by valves and a pair of contractile vessels occurs in the posterior abdominal regions of some ratites and Anseriformes. True lymph nodes are of limited occurrence, two pairs occur in some Anseriformes, but solitary lymphatic nodules are widespread.

ii The thymus consists of irregular lobes in the cervical region and achieves its maximal size at sexual maturity. Involution then occurs. A dorsal medial diverticulum of the proctodeum, the cloacal bursa, is unique to birds. Its developmental history is similar to that of the thymus, despite their very disparate locations, and both are the primary sites of lymphocyte development. The thymus is involved, as in mammals, in cellular immune responses but the bursa is the principal site of antibody synthesis.

SPECIAL SENSE ORGANS

(a) The eyes

i Vision is an extremely important sensory modality for all birds and their eyes are extremely large. Those of eagles have similar dimensions to those of humans despite the huge disparity in body size. The overall shape (Fig. 8.11) varies from flat, through globular, to the rather tubular structures in owls, but the detailed organization is closely reminiscent of that in diurnal lizards.

ii The retina is thicker than that of most mammals and lacks blood vessels. Diurnal birds have more cone cells than rods and the opposite can be true of nocturnal species. The former have colour vision. One or more

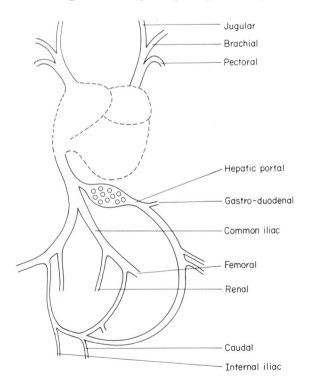

Fig. 8.10 Stylized diagram of the venous system of a bird. The proximal component of the brachial venous system is called the subclavian vein.

Jugular
Brachial
Pectoral
Hepatic portal
Gastro-duodenal
Common iliac
Femoral
Renal
Caudal
Internal iliac

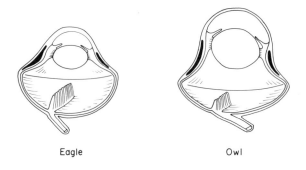

Fig. 8.11 Diagrams showing the differing shape of the eye in eagles and owls. (After Walls.)

regions of maximum discrimination have particularly high concentrations of cones and these *areas* may contain a fovea comparable with that of mammals. Three principal types of *areas* and foveae occur:

graminivorous species have a single *central area* that is situated close to the optic axis and may have a fovea;

aquatic birds such as waders have a ribbon-like

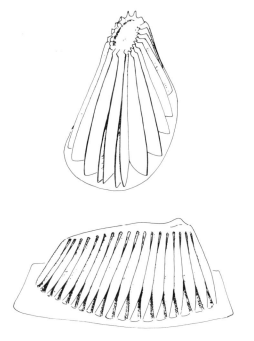

Fig. 8.12 Two contrasting types of pecten from the eye of birds. Above: the vaned type of the ostrich. Below: the pleated type with thick vertical folds that occurs in carinates.

central area and their heads are normally held so that this lies close to the horizontal plane;

humming birds and falcons have two foveate areas. A central one lies close to the optic axis and a second lies laterally. These underlie the bird's ability to pursue prey or feed on the wing.

iii Nutrients reach the retina via the pecten oculi. This structure projects into the vitreous humour at the blind spot where the optic nerve leaves the eye. Unlike the cone-shaped structure of lizards, the pecten of most birds is folded and has a much greater surface area. Two principal types occur. A vaned form characterizes the large cursorial ratites whilst a pleated form characterizes most carinates (Fig. 8.12). In general the structure is largest and most prominent in diurnal predators. This supports the contention that, besides its nutrient capacity, it also assists in visual discrimination by breaking up the field of vision.

iv The retinae of many birds contain oil droplets. These are brightly coloured in diurnal species and either pale yellow or colourless in nocturnal ones. They probably assist in colour vision by acting as intra-ocular filters.

(b) The ear

i Birds also have an extremely well-developed sense of hearing. Most birds use sounds as instruments for locating members of their own or other species but directional localization is particularly impressive in the echolocating swiftlets or those owls with asymmetrical ears.

ii The external ear is a relatively short tube that terminates internally at the tympanic membrane. Externally the presence of special covert feathers reduces both drag and wind-generated noise. In the middle ear the vibrations of the tympanic membrane are transmitted to the perilymph of the inner ear by means of a lateral extracolumellar cartilage and the medial bony columella.

iii The avian cochlea is a short and more or less straight tube that ends in the lagena. The cochlear duct, containing endolymph, is separated from the occluded scala vestibuli by a thick and vascular tegmentum vasculosum, and from the well-developed scala tympani by the basilar membrane (Fig. 8.13). It is the blind apical region of the cochlear duct that forms the lagena. The

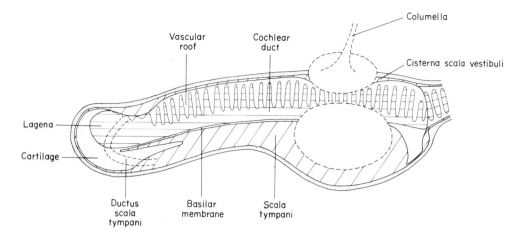

Fig. 8.13 The organization of the cochlea in birds.

concentration of sensory cells in the papilla basilaris is about ten times greater than that in the mammalian organ of Corti. Sensory fibres in the auditory nerve run from these cells to the angular and magnocellular nuclei in the medulla oblongata.

iv Membranous semicircular ducts lie within their bony canals and originate from the sac-shaped utricle. At their origin each has an ampullary enlargement. The vestibular sensory areas, which are of great importance to the animals during flight, comprise cristae in the ampullae together with the macula utriculi and macula neglecta of the utricle, and the macula sacculi of the saccule. The cristae of both the anterior and posterior ducts bear a horizontal fold, the cruciate septum, which lacks sensory epithelium but divides that of the rest of the organ into two regions. Such structures are not limited to birds but also occur in some scansorial and aquatic mammals.

The sensory hairs in the macula utriculi and macula sacculi are embedded into an otolithic membrane which contains calcareous structures. Movements of the endolymph displace the membrane and systematically distort the neuroepithelial cells.

THE CENTRAL NERVOUS SYSTEM

(a) The spinal cord

The variation in vertebral number which was noted above is paralleled by variations in the number of segmental elements within the spinal cord (cf. Table 8.1). The ventral units are usually larger than the

Table 8.1 The segmental composition of the spinal cord in two contrasting genera.

	Cervical segments	Thoracic segments	Lumbosacral segments	Coccygeal segments
Struthio	15	8	19	9
Columba	12	8	12	6

dorsal ones and this discrepancy is reflected in the relative sizes of the ventral and dorsal horns. Marked enlargements occur in the cervical and lumbar regions in association with the innervation of wings and legs. The lumbar enlargement predominates in the large cursorial genera, the cervical one in volant forms. Within the cord the majority of axons are short and only traverse a few segments.

(b) The medulla oblongata

i As in other classes the medulla is a heterogeneous structure including both ascending and descending fibre bundles, the nuclei at which cranial nerves V–XII originate or terminate, and others which comprise local aggregations of the reticular formation (Fig. 8.14). XII arises from two hypoglossal nuclei which both represent anterior prolongations of the ventral horn in the spinal cord. They are somatomotor components supplying postotic somites which contribute to the musculature of tongue and trachea. X has a dorsal motor nucleus which is implicated in cardio-inhibition. The majority of afferent fibres within both the vagal and glossopharyn-

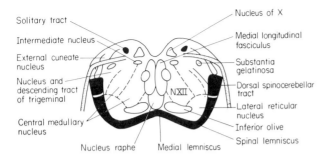

Fig. 8.14 Outline of the organization of an avian medulla.

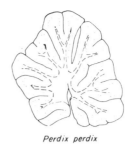

Perdix perdix   *Meleagris gallopavo*

Fig. 8.15 Longitudinal sections of the cerebella of a partridge and turkey showing the greater degree of subfoliation in the latter, larger bird. (After Senglaub.)

geal nerves pass to the nucleus of the solitary tract. The facial (VII) motor aggregation has dorsal, intermediate and ventral components and is closely apposed to that of the trigeminal (V). Trigeminal afferents pass to both the main sensory trigeminal nucleus and the more anterior mesencephalic sensory nucleus which is particularly implicated in the control of bill opening.

ii Fibres originating in the cochlear region, and passing within the auditory nerve (VIII), terminate in the angular and magnocellular nuclei where they maintain a tonotopic distribution. These nuclei, together with the neighbouring laminar nucleus, are greatly enlarged in those owls with a well-developed directional localization ability. Vestibular projections terminate in six principal vestibular aggregations.

iii The inferior olive is a rather massive association of lamellae which project to the cerebellum. In addition rather poorly developed pontine nuclei parallel the more elaborate structures of mammals.

(c) The cerebellum

i All avian cerebella possess many more lobules than do those of crocodiles. During development, two clefts, the primary and secondary fissures, divide the cerebellum into three main lobes which are closely coincident with those of crocodiles. However, development continues, to give the ten principal lobules of the adult, and the final number of sublobular components is related to overall body size (Fig. 8.15).

ii Within the cerebellum lie the cerebellar nuclei. Three main ones are usually designated as the internal, intermediate and lateral units of each side. However, various workers have emphasized their essential continuity.

iii Ascending projections to the cerebellum are of sensory origin. Proprioceptive fibres passing in the spino-cerebellar tracts. Efferent projections are known to pass to the reticular formation and red nucleus. It is assumed that the overall controls exerted by the organ are comparable with those of other classes (cf. page 150).

(d) The mesencephalon

Cranial nerves III and IV continue to have their nuclei in the midbrain. The most obvious mesencephalic component is, however, once again the optic tectum. As in diurnal lizards, etc., this has fourteen laminae that are usually referred to six main strata. The lateral mesencephalic nucleus is the probable homologue of the torus semicircularis in amphibians and the inferior colliculus in mammals. It receives projections from the medullary auditory nuclei via the lateral lemniscus and is particularly large in the echolocating swiftlets. The isthmo-optic nucleus, lying close to the oculomotor and trochlear aggregations, is involved in visual feedback systems, and the red nucleus is the source of the rubrospinal motor tract.

(e) The diencephalon

i The epithalamus includes both the habenular nuclei and bears the pineal organ. Although some of the cells of this last resemble definitive photoreceptors in the young, the adult organ is predominantly glandular and secretes melatonin. This is involved in the control of both circadian rhythms and the rapid ovarian maturation that precedes laying.

ii Two ascending visual pathways have been identified and both of these have relay stations in the thalamic regions. The more direct one involves projections from the retina to the dorsal thalamic regions, and from there to the Wulst (q.v.). The less direct one is a retino-tecto-rotundo-ectostriate pathway. The rotund nucleus is one of the largest thalamic nuclei. Another prominent thalamic aggregation is a diencephalic relay on the ascending auditory pathway. This is the ovoidal nucleus which receives projections from the lateral mesencephalic nucleus and projects to the neostriatum.

iii Three principal hypothalamic nuclei contribute secretory fibres to the hypothalamo-hypophyseal tract. These are the paraventricular, supraoptic and infundibular cell aggregations. The hypothalamus is continuous, via the tuber cinereum, with the median eminence. Blood vessels running through particular regions pass to different foci in the adenohypophysis.

(f) The cerebral hemispheres

i The avian cerebral hemispheres are solid structures and therefore contrast with the cortical structures of mammals. It is, however, clear that the intermediate striatal aggregations and the Wulst are derived during development from cells that migrate inwards from the general pallium. They are therefore, in broad terms, the homologues of mammalian neocortex.

ii The paleostriatum primitivum and the paleostriatum augmentatum are the deepest units and the probable homologues of mammalian basal ganglia. All the fibre projections which enter the hemispheres pass through them.

iii The intermediate striatal aggregations lie above, or in front of, the paleostriatal moieties (Fig. 8.16). They are now known to include fields of cells with very specific sensory and motor involvements. The archistriatum is the most anterior station on somatosensory pathways. The neostriatum includes an anterior auditory projection area, and the ectostriatum is the most anterior station on the tecto-rotundo-ectostriate ascending visual pathway. In addition the archistriatum is implicated in the control of aggressive behaviour and lay, whilst stimulation of the neostriatum can elicit feeding.

iv The dorsal and accessory hyperstriatal aggregations together constitute the Wulst or sagittal eminence and lie above, or in front of, the ventral hyperstriatum and other intermediate striatal aggregations. They include anterior auditory fields and the terminal stations of the

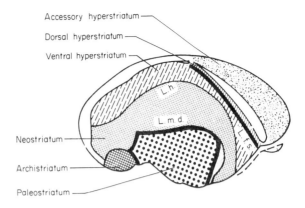

Fig. 8.16 Diagram of the hemisphere of a crow. In some other birds the dorsal and accessory hyperstriatum are dorsal. L.h., hyperstriatic fibre làmina; L.m.d., dorsal medullary fibre lamina.

ascending visual pathway that passes through dorsal thalamic regions.

v Cortical structures are best developed in emus, cassowaries, kiwis, etc. They are greatly reduced amongst carinates where they are better developed in the young than in the adults, and more extensive in aquatic genera than in others. In general terms they comprise archipallial and paleopallial elements giving hippocampal, parahippocampal, periamygdalar and prepyriform fields.

(g) The olfactory bulbs

All birds have a sense of smell but the size of the olfactory bulbs varies immensely. Their maximal development occurs, not surprisingly, in the kiwi, and amongst carinates they parallel the cortical components of the cerebral hemispheres and are most developed in aquatic genera.

BEHAVIOUR

i Many aspects of avian behaviour are closely related to the type of development of the young. Broadly speaking, nidifugous behaviour is reminiscent of the reptilian condition and occurs in, for example, many Ratites, game birds, cranes, jacanas, bustards, divers, grebes, ducks and geese. It occurs in its most extreme form in the Megapodes. For example, in *Megacephalum maleo* the eggs are laid in the humus floor of the forest

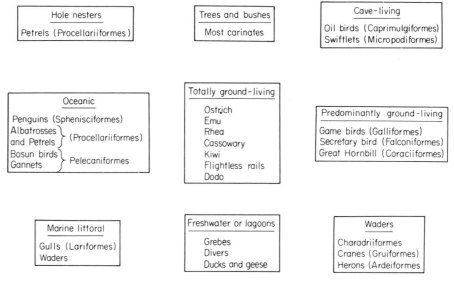

Fig. 8.17 The adaptive radiation of birds as exemplified by certain species or orders.

and the young are as independent of the parents as are those of many reptiles.

Nidicolous behaviour in which the young are 'helpless' nestlings relying on their parents for food and warmth, typifies pelicans, petrels, pigeons, herons, falcons, owls, cuckoos, woodpeckers and the songbirds. An intermediate condition marks the kiwis and flamingoes. Although the young of gulls and penguins can leave the nest at an early stage they remain dependent on the parents for food.

ii Imprinting, in which the young follow and imprint upon the first moving object they see during a critical sensitive period, was initially described in nidifugous birds. It is clear that a similar sensitive period also occurs in nidicolous species but at a later period after hatching.

iii The sequence of events which characterizes parental behaviour is intimately related to hormonal levels. Specific components of the song of the other sex stimulate maturation of the gonads in many species. In male pigeons the initial phases of courtship, where aggressive components are much in evidence, depends upon high follicular stimulating hormone/androgen titres. A subsequent phase of high oestrogen levels leads to nest demonstration. FSH leads to nest building and progesterone secretion to incubation.

iv Both sexes share the incubation of eggs in many species but the male does it in some phalaropes. A wide variety of vocal and visual displays are associated with the acquisition of, the defence of, and the solicitation of females into, territories within which the nest or nests are constructed. Although many birds are monogamous, at least for a particular season, others, wrens, for example, are polygamous, and in some species various males that are close kin of the principal male assist in rearing the young.

THE ARCHAEORNITHES

The morphological characteristics of the Jurassic genus *Archaeopteryx* are universally interpreted as representing a condition which is intermediate between those of the ancestral coelurosaur stock on the one hand and more recent avian genera on the other (Fig. 8.18). In many respects they resemble small dinosaurs but the presence of feathers necessarily associates them with birds. This similarity to reptiles even extends to what we can deduce about their brains.

THE RATITE ORDERS (Fig. 8.19)

The large cursorial birds are usually placed in separate orders. The ostriches (Struthioniformes), rheas (Rheiformes), cassowaries and emus (Casuariiformes),

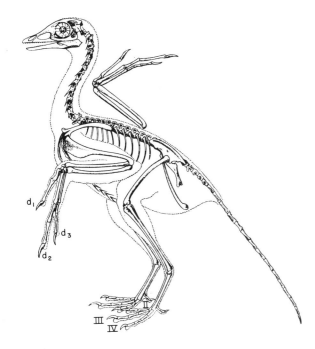

Fig. 8.18 The skeleton of *Archaeopteryx*.

together with the kiwis (Apterygiformes) all differ from the carinate birds in the absence of a carina or keel to the sternum. Their clavicles are rudimentary, or absent, and the hind vertebrae are not fused into a pygostyle. Their paleognathous palate is similar to that of the Tinamiformes amongst the carinates whilst serum protein tests suggest that their closest affinities are with both these and the Galliformes. Their own interrelationships remain controversial. During the Tertiary period two other families of flightless birds existed. *Diatryma* is a large, big-headed genus from the Lower Eocene of Wyoming which stood some seven feet high in life and *Aepyornis* inhabited Madagascar. The moas from the Quaternary deposits of New Zealand included a number of species one of which was eleven feet tall. Their final extinction appears to have been the result of human activity.

THE IMPENNAE

i   The penguins are members of the order Sphenisciformes and are separated by some authorities from all other orders. They are flightless oceanic birds whose wings are modified as paddles and whose legs are set well back. As a result of this last factor they are the only birds that walk erect. It has been suggested that they are primarily flightless on the grounds that their bones are not pneumatic, the rectus abdominis muscles are primitive and their metapodials resemble those of cursorial dinosaurs. However, most people view them as highly specialized carinates.

ii   At the surface they swim low in the water and the colouration of the exposed head and nape is important in sexual display. They feed on cephalopods and crustaceans. Their principal enemies are the leopard seal and skuas. The jackass penguin has been farmed for its eggs.

iii   Two species occur on the antarctic continent, four on ice-covered water, six are south temperate and four occur in the southern tropics. The isotherm for the 20° C annual air temperature is a rough guide to their northern limits. In broad terms they stop breeding at the latitude at which turtles start to do so. Adaptive features include feathered nares in those species breeding in the snow. *Pygoscelis adelea*, which is the only strictly antarctic member of its genus, has feathered nares. *P. papua*, the gentoo, lacks this feathering and *P. antarctica* exhibits an intermediate condition. The king and emperor penguins are of large size, have small extremities and huddle together during blizzards. Species of small body size, like the adélie and gentoo, have short incubation periods of around 35 days and the chicks develop in 5–8 weeks. In *Aptenodytes*, the emperor, there is a compromise between the necessarily long developmental and growth period and the short feeding period. Eggs are laid in June, one month after the sea ice has formed. The male parent incubates them for two months whilst the females are at sea. He regurgitates food and gives the chick its first meal. In August the females return and take over. By September the feeding grounds are moving nearer and the young themselves reach the breaking ice in January.

THE PROCELLARIIFORMES

The remaining orders exemplify the wide diversity of the carinate birds. The procellariiform genera are oceanic birds characterized by having their external nostrils at the end of a tube (Fig. 8.20) from which oil can be ejected. The bill retains several relatively discrete horny plates but, these features aside, there is considerable convergence with the gulls (but see Table 8.2). The giant fulmar which gorges carrion or the eggs and young of penguins is the only species that habitually feeds on

Kiwi

Ostrich

Cassowary

Fig. 8.19 The diversity of ratite birds.

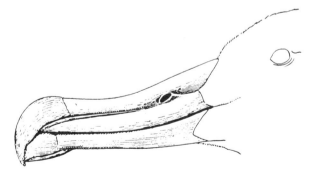

Fig. 8.20 The beak of an albatross showing one of the large nostrils.

Table 8.2 Contrasting characteristics of the petrels and gulls.

Procellariiformes	Lariformes
Only one immaculate egg	2–3 cryptic eggs
Eggs laid in burrows or on screes	Eggs laid in the open
Unable to walk	Walkers
Nidicolous young	Nidifugous young
Naked young	Downy young

land. *Hydrobates* feeds at sea by submergence whilst the diving petrels dive and swim under water. *Pachyptila*, the whale bird, has a filter feeding system using bristle feathers to strain food from water in its mouth.

THE PELECANIFORMES

i The pelicans and their allies are characterized by being aquatic birds in which the hallux is turned in and connected to the other toes by a web giving the totipalmate condition. This web is somewhat emarginated in the frigate birds. The sutures of the bill components are also clear as in the order Procellariiformes.

ii In *Phaethontes*, the tropic birds, the middle pair of rectrices are long and attenuated. Pelicans, gannets, darters, cormorants and frigate birds also all have characteristic tail shapes which can be used for identification. Pelicans habitually fish in formation and drive their prey into the shallows where they are more easily captured. Cormorants have, in the past, featured large in the peasant fisheries of China and Japan. The fishermen fitted collars round their necks to prevent them swallowing their prey and held them on the end of leashes.

THE COLYMBIFORMES AND PODICIPITIFORMES

These orders include the foot propelled divers in which the tibial part of the leg is bound to the body by muscles. The epigamic displays are mutual and include the well-known plesiosaur dance. Some inconvenience awaits the unwary due to the contrasting nomenclature used by authors on each side of the Atlantic. This is summarized in Table 8.3. The differences between the two types of bird are also summarized in Table 8.4.

Table 8.3 A comparison of the differing nomenclature on each side of the Atlantic.

Wetmore's nomenclature	British nomenclature	Common names
Gaviiformes	Colymbiformes	Divers or loons
Colymbiformes	Podicipitiformes	Grebes

THE ARDEIFORMES (≡ Ciconiiformes)

The herons, night herons and egrets differ from the related storks and ibises in the possession of pectinate middle claws. These are used, in association with keratin from the powder-down patches, during preening. The middle toes of storks and ibises are generally short and the birds are less aquatic than herons.

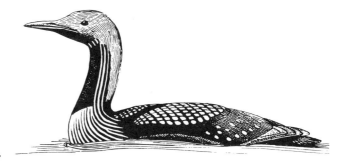

Fig. 8.21 A diver.

Table 8.4 Some differing characters of divers and grebes (cf. Fig. 8.21).

Divers	Grebes
Front digits webbed	Digits lobed
Tail feathers normal	Tail feathers absent
Cervical vertebrae 14–15	Cervical vertebrae 17–21
Thoracic vertebrae free	Thoracic vertebrae 1–4 fused
Sternum short	Sternum long
11 primary feathers	7 primary feathers
2 cryptic eggs	4–6 immaculate eggs
Marine island nesters	Freshwater floating nests

THE PHOENICOPTERIFORMES

The flamingoes are characterized by their highly mobile upper mandible which is used for filter feeding. Their exact taxonomic relationships remain somewhat controversial although they were placed alongside the herons by Wetmore.

THE GALLIFORMES

i The game birds are cosmopolitan in distribution although absent from Micronesia. They are short-winged and have well-developed feet armed with strong claws which they use to scratch for food. As mentioned above they are the group of carinates which, together with the Tinamiformes, appear from their proteins to be most closely allied to the ostriches, emus, etc.

ii The hoatzin, *Opisthocomus*, occupies a rather isolated taxonomic position. It has certain affinities with the cuckoos and is one of the few birds to have 'eyelashes' (cf. hornbills). The nidifugous young possess

Fig. 8.22 A hoatzin.

a clawed pollux and clawed index digits which are used whilst clambering through the branches of the mangrove swamps in which they live (Fig. 8.22).

THE GRUIFORMES

The cranes are a rather heterogeneous assemblage of cursorial or wading birds associated by some authors with the Charadriiformes or definitive waders. They have two outstanding characteristics. They lack a crop and the hallux is elevated and tends to be reduced. There is a consequential difficulty in perching. Occurring widely but not in South America they can be distinguished from herons, when flying, by their outstretched necks. Herons have their necks doubled back when in flight. Other representatives include the coots, gallinules and rails.

THE ANSERIFORMES

i The tree-ducks, geese, swans and ducks are envisaged as descendants from forms similar to the Anhimae or screamers (Fig. 8.23). In the Anserines, tree-ducks, swans and geese, the tarsus is reticulated, there is one moult a year and the sexes are alike.

Fig. 8.23 *Chauna torquata*, a member of the Anhimae.

ii In the Anatinae the tarsi are scutellate and there are two moults per year. The Anatini or dabbling ducks are surface feeders and leap out of the water at take-off. The Anthyini or pochards have a deep lobe on the toes. The sea-ducks or mergansers form a third group—the Mergini.

iii Down is used for nest-building and the eggs are frequently covered when the parent leaves the nest (cf. grebes which use leaves). Large and invulnerable species have white down (e.g. swans). Smaller and more vulnerable species like the mallard, wigeon and pintail, have dark down. Small vulnerable species with white down, such as goldeneye, smew and goosander, nest in holes.

THE FALCONIFORMES

i The diurnal birds of prey fall into three types. The Sagartiidae or secretary birds differ from the rest in their overall appearance and their long scaly legs which protect them from their snake prey.

ii The Cathartae, condors or New World vultures, differ from the Falconae in their scaled feet which are not adapted for grasping. The digits can bear an interdigital membrane and the short claws are reminiscent of those of game birds. Turkey vultures which scavenge around Brazilian slaughter houses have long been protected for their contribution to hygiene.

iii The Falconae includes all other diurnal birds of prey. Their feathers have an after-shaft which is lacking in Cathartae. The Accipitridae (hawks, ospreys and eagles) and the Falconidae (falcons and caracaras) have rather specific habitats and food. Nearly all capture their prey on the wing but the techniques vary greatly. Species that are characteristically found in open country have broad wings and are good soarers. The golden eagle kills by a swift attack, the buzzard and harriers by a low flight and sudden pounce. The goshawk, a bird of woodland, follows its prey from a perch and then pounces. The sparrow-hawk flies along hedges. Stooping is characteristically associated with narrow pointed wings which are partially closed during the stoop. Speed is of paramount importance when the hobby catches dragonflies or the merlin catches small birds. Hovering followed by a sudden drop characterizes the kestrel (cf. some African kingfishers), whilst the osprey plunges and catches fishes in its talons. The honey buzzards dig out the nests of bees and the relevant excavating movements are exhibited by nestlings at an early stage. Vultures scavenge.

THE CHARADRIIFORMES

Some authors group the definitive waders with the gulls and auks. They are, *par excellence*, the birds of marine estuaries and many are arctic-alpine nesters. The plovers hunt by audio-visual methods, whilst redshank, knot, sanderling, dunlin, etc., catch their prey by probing in the mud. The oystercatcher, in particular, has a bill that is adapted for opening bivalve shells.

THE LARIFORMES

The gulls are characteristic species of the ocean littoral. At first sight they may be confused with some procellariiform species (see Table 8.2, page 124). The Rhynchopidae, skimmers, which have a narrow beak with a long lower mandible (Fig. 8.24) are unmistakable. They fly along over the water surface with the lower mandible submerged and snap the bill closed if it comes into contact with a fish. Their flight over the surface can itself act as a lure to surface-feeding fishes and they frequently backtrack over the same flight path thereby catching any fish attracted to the surface.

THE ALCIFORMES

Those authors who place the gulls in the order Charadriiformes do the same with the auks. In addition

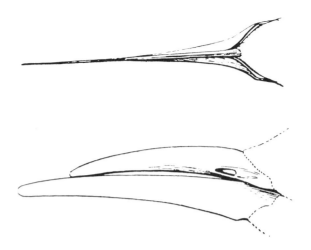

Fig. 8.24 Ventral and lateral views of the beak of *Rhynchops*, the skimmer.

to the great auk or garefowl, extinct since 1844, this order includes the little auk, the razor-billed auks, the guillemots and the puffins. They are marine, diving birds with elongated bodies and stubby necks. Their feet, which have only three toes, are webbed, large, and set well back near the base of the tail. This both increases the power of their swimming strokes and gives them an upright stance similar to that of penguins. They usually breed on steep shores or cliffs and there is usually a single egg.

THE COLUMBIFORMES

The pigeons and doves differ from their relatives the sand-grouse in the characters summarized by Table 8.5.

Table 8.5 Comparison between the pigeons and doves (*Columbi*) and the sand grouse (*Pterocletes*), both groups are members of the order Columbiformes.

Columbi	Pterocletes
Possess a cere	Lack a cere
Tarsi rarely plumed	Tarsi plumed
Construct a nest	Eggs laid in depression on ground
1–2 immaculate eggs	3 cryptic eggs
Nidicolous young	Nidifugous young

THE MUSOPHAGIFORMES

The plantain-eaters have sometimes been associated with the cuckoos because their fourth toe can be rotated to lie in front or behind the foot. This semi-zygodactylous condition, therefore, recalls the zygodactyly of cuckoos. Their downy young contrast with the naked nestlings of cuckoos and their feathers contain a unique copper-based pigment turacin which is soluble in alkaline fluids. It is synthesized, no matter how low the copper concentration in the body, six weeks after hatching. All other green or greenish birds have structural colours.

THE CUCULIFORMES

The roadrunners, Crotophaginae, and cuckoos together constitute an order which is cosmopolitan outside the polar regions. The roadrunners make communal nests and the cuckoos include the best-known brood parasites. The females of the European cuckoo have territories and each lays eggs of one type only which resemble those of the host species in both size and colouration. The Asiatic *Cuculus poliocephalus* has various egg types. Those laid in the nests of *Phylloscopus*, for example, are white; others are chocolate coloured.

THE STRIGIFORMES

The owls or nocturnal birds of prey have large, forward-facing eyes, a very pneumatic skull and soft fluffy plumage. The ear openings are larger in species that inhabit high latitudes than in those inhabiting the tropics. This reduces the interference provided by the vast number of insect and other noises in low latitudes since the owls hunt predominantly by hearing. They can localize sounds of around 15 kHz with great precision. This is assisted in certain genera by the asymmetrical form of the ears on the two sides. Modifications of the feathers (Fig. 8.25) prevent turbulence and facilitate the

Fig. 8.25 An owl feather showing modifications for silent flight.

silent flight. Species such as the snowy owl necessarily hunt in daylight during the polar summer and most also fly at twilight.

THE CAPRIMULGIFORMES

Oil-birds, whip-poor-wills and nightjars are another essentially nocturnal assemblage. They provide an interesting comparison with the bats amongst mammals. The order includes the only birds which regularly enter a state of hypothermic torpor and, together with the cave-swiftlets (Micropodiformes), those that use echolocation. *Steatornis*, the oil-bird, emits numerous sounds including short sharp ones of 7 kHz and 1/2000 second duration. It is these that are used in echolocation and their characteristics are closely similar to those used by *Rousettus*, the only megachiropteran (q.v.) that echolocates. The birds can fly in the dark without collisions. If the ears are plugged with wax they lose this ability.

THE MICROPODIFORMES

The swifts are predominantly aerial and insectivorous. Their short legs render them rather helpless on the ground. They have pointed wings and achieve speeds that are well in excess of $100 \, \text{km h}^{-1}$. The cave swiftlets of South East Asia build their nests in caves and use echolocation. They show, therefore, convergent evolution with both the oil-bird and bats. The nests of some are made from buccal secretions and form the basis of bird's nest soup.

THE TROCHILIFORMES

The humming-birds are again a very homogeneous order. Their generally small size, long bills and elaborate tail feathers render them quite unmistakable (Fig. 8.26). Exclusive to the New World, some 450 species and subspecies are known of which those of the genus *Archilocus* are minute.

THE CORACIIFORMES

The kingfishers, bee-eaters, hoopoes and hornbills are a far more heterogeneous assemblage. The hornbills, medium-sized or large birds inhabiting Asia and Africa, are renowned for both their large bills and their breeding habits. Once courtship is over the female retires to a hollow tree and seals herself in with an adobe-like material made up of dung and pellets of mud gathered

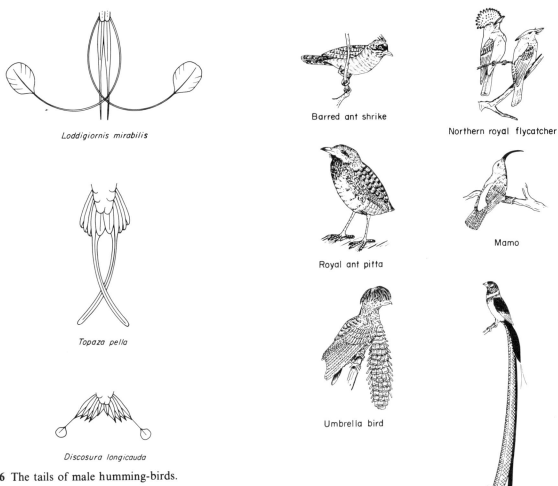

Fig. 8.26 The tails of male humming-birds.

Loddigiornis mirabilis

Topaza pella

Discosura longicauda

Barred ant shrike

Northern royal flycatcher

Royal ant pitta

Mamo

Umbrella bird

Steganura

Fig. 8.27 Varied passerines.

by the male from the forest floor. He passes them to her in the form of saliva-permeated pellets. The bill exhibits convergent evolution with those of toucans (Rhamphastidae) amongst the Piciformes. The rhinoceros hornbill is the largest species. Unlike the tail feathers of other birds the two central ones are not moulted together but instead are replaced alternately so that one is always present perhaps acting as a rudder.

THE PICIFORMES

The woodpeckers, honey guides, puffbirds and toucans are again a heterogeneous assemblage. The unusual structures associated with the tongue in woodpeckers have already been referred to. The 37 species of toucans live in Central and South America. The curl-crested aracari is perhaps the most sensational as it has curled and rumpled cellophane-like crown feathers. The stereotyped behaviour of *Indicator indicator*, the honey-guide, perhaps evolved in association with mammals such as the honey badgers. It points out the nests of wild bees and is used for this purpose by man.

THE PSITTACIFORMES

The parrots, parakeets, etc., are highly specialized perching birds whose skull structure distinguishes them from other groups (cf. page 111). Best known are the sulphur crested cockatoos, the macaws and the budge-

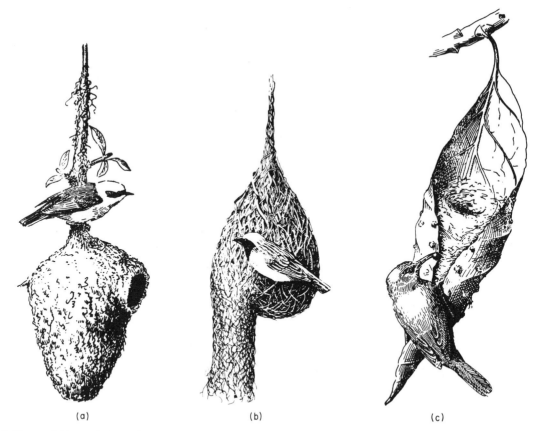

Fig. 8.28 Nests of (a) penduline tit, (b) ploceid finch, and (c) tailor bird.

rigars that have long been kept by aviarists. The kea parrot of New Zealand (*Nestor notabilis*) achieved notoriety when it began to rip open the back of sheep but the owl parrot, *Strigops*, another inhabitant of New Zealand, is perhaps more interesting. It cannot fly and is strictly ground dwelling. The structural colours of macaws are striking and apt to change when the feathers are wet.

THE PASSERIFORMES

The perching birds and song-birds (Fig. 8.27) are the most diverse avian order. The family Corvidae (ravens, crows, jays, magpies, etc.) have long been known to be highly intelligent and have been made the subject of many behavioural studies. In fact the learning abilities of many birds have long been undervalued. The birds of paradise are equally renowned for their elaborate courtship behaviour. The nests of many species of passerines are also elaborate. Notable in this respect are those of

tailor-birds, weaver birds, oven-birds and various titmice (Fig. 8.28).

FURTHER READING

Any library will provide numerous texts on ornithology and particular groups of birds.

FARNER D.S. & KING J.R. (eds) (1969) *Avian Biology*, 5 vols Academic Press, London.

KING A.S. & McLELLAND J. (1975) *Outlines of avian anatomy*. Baillière Tindall, London.

MATTHEWS G.V.T. (1968) *Bird navigation*. Cambridge University Press, Cambridge.

Nomina anatomica avium (1980). Baumel J. J. (ed.). Academic Press, London.

THOMSON SIR A. LANDSBOROUGH (ed.) (1964) *A new dictionary of birds*. Nelson.

YAPP W.B. (1970) *The life and organisation of birds*. Edward Arnold, London.

WEBB J.E., WALLWORK J.A. & ELGOOD J.H. (1979) *A guide to living birds*. MacMillan, London.

9 · The Mammals

INTRODUCTION

The outstanding features of mammals are well known. They are endothermal amniotes which have an external covering of hair and feed their young on milk. Permo-Triassic fossils are sometimes difficult to ascribe to either reptiles or mammals but the presence of a single bone in the lower jaw is certainly taken as indicating definitive mammalian status. In general terms, the mammalian characters are heterodonty, diphyodonty, multicusped and multirooted teeth. Additional jaw elements of reptiles are incorporated into the middle ear. Most of their distinctive features are related to the evolution of increased activity and greater care of their young. They clearly resemble birds in this but the two groups achieved endothermy and parental care quite independently.

INTEGUMENT

i The skin is again extremely important and contributes to protection, thermoregulation, exteroception and water regulation. It consists of a covering epithelium, the epidermis, overlying the dermis and the hypodermis or subcutis (Fig. 9.1). The epidermis is a stratified, squamous, keratinized epithelium comprising a basal stratum germinativum; a stratum spinosum with lighter flatter cells; a stratum granulosum; a stratum lucidum, and, finally, a stratum corneum consisting of several layers of flat cells that lack nuclei.

ii Under the electron microscope the cells of the stratum germinativum can be seen to be rich in polysomes, those of the s. spinosum have numerous tonofilaments which are arranged in an orderly manner and end in desmosomes, and those of the s. lucidum and s. corneum are devoid of organelles. Melanin is contained in melanocytes, which also occur in the dermis, and is synthesized in melanosomes which are elongated structures of neural crest origin containing an internal lamellar organization and periodic striations.

iii The dermis consists of reticular and papillary layers, and merges with an underlying hypodermis. In some places, in the metacarpus of the Equidae for example, the hypodermis is merely a thin layer of dense connective tissue which fuses with the periosteum but elsewhere it can be a thick layer of fat which acts as a protective cushion.

iv Both sebaceous and sweat (sudoriferous) glands are widespread. The former are generally associated with hairs but this is not the case with certain specialized structures such as the submental gland of the cat, the infra-orbital and inguinal glands of some ruminants, etc. Two principal types of sweat gland occur. Merocrine glands are mainly limited to areas like the foot, pad or hoof, which have few or no hairs, whilst apocrine glands are associated with hairs and are, therefore, more widely distributed. They include the axillary glands of humans which secrete pheromones.

v The mammary glands are the most striking modifications of sweat and sebaceous glands and contribute to one of the principal mammalian characteristics—

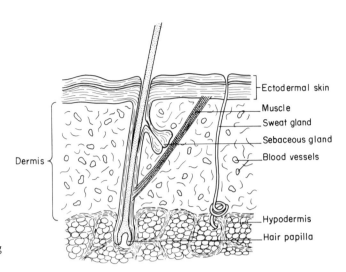

Fig. 9.1 Diagram showing mammalian skin and hair.

feeding the young on milk. A diffuse bed of glands secretes milk on to the venter of female *Ornithorhynchus* and this is then lapped up by the young. In contrast, the discrete structures within the pouch of marsupials, and on the venter of placentals, are suckled by the young. The presence of young, at widely different stages of

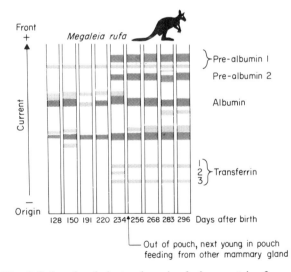

Fig. 9.2 Starch gel electrophoresis of whey proteins from milk of a kangaroo during the second half of lactation showing the change of composition just prior to the suckling leaving the pouch and the next young entering it. The mother then produces early and late milk simultaneously from adjacent mammary glands. (After Bailey & Lemon.)

development, in the pouch and at the heel of kangaroos is associated with quite different milk content in different mammae. These differences relate, in particular, to the concentration of transferrins which are the source of iron and copper for the marsupial neonate. In placentals the fetus obtains these elements whilst still in the womb but in marsupials high levels are needed during the early period of pouch life (Fig. 9.2).

vi Hairs are another primary mammalian characteristic and arise from epidermal invaginations—hair follicles—into which the hair papilla protrudes centrally. Peripheral cells produce the internal root sheath which surrounds the hair root below the level of the sebaceous glands. Hairs themselves comprise an outer cuticle, a cortex of cornified cells whose tonofibrils give the hair rigidity, and a pigmented medulla containing trichohyalin granules. The overall shape of the hair can be very characteristic as in the Chiroptera (Fig. 9.3).

vii Sensory nerve endings, Merckel's discs (pressure), Meissner's corpuscles (pressure) and Krause's end bulbs (cold) are widely distributed sense organs. Similar structures occur in birds.

viii Various specialized structures also exist. The hooves of the Equidae are actually modified skin and comprise a wall, sole and frog. The bars, internal continuations of the wall, and the frog are absent from the analogous claw structures of Artiodactyla. In terrestrial carnivores the sole is enclosed by the wall and soft horn.

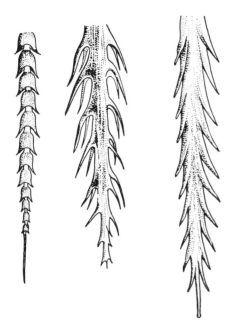

Fig. 9.3 The appearance of bat hairs.

SKELETAL SYSTEM

(a) The skull

i   The skulls of mammals (Figs 9.4 and 9.5) show close affinities with those of synapsid reptiles and only differ from those of tritylodonts and ictidosaurs in their enlarged braincase; in the presence of a single bone, the dentary, in the lower jaw; and in the incorporation of the angular and articular bones into the auditory region. The angular becomes the tympanic bone, the articular the malleus, and the quadrate is associated with them as it is transformed into the incus. Interpretation of the mammalian skull is therefore facilitated if the reptilian ground plan is kept in mind.

ii   In dorsal view the principal characteristic is, as in birds, the swollen braincase. The temporal arches stand out from the sides of the skull and there is the normal tetrapod series of parietals, frontals, nasals and pre-maxillae. No bar occurs behind the orbit in 'primitive' mammals and the reptilian postorbital is lost, as are the pre- and post-frontals. The lachrymal is absent from monotremes, is small, and confined to the orbital region, in placentals. The nasalia project freely over the nares and meet the premaxillae on either side. The jugal lies beneath the orbit and contributes to the zygomatic arch, whilst the squamosal forms both the hind end of this arch and also the cavity for the jaw articulation. As in reptiles there is considerable fusion of the occipital elements. The basioccipital, which lies below the foramen magnum, is fused with the exoccipitals which contribute in large part to the paired condyles, and during embryonic development a post-parietal fuses with the supraoccipital. A transverse lambdoidal crest providing an attachment for neck muscles is often present across the upper margin of the occiput.

iii   On the underside of the braincase basisphenoids lie in front of the basioccipital, are bounded laterally by the

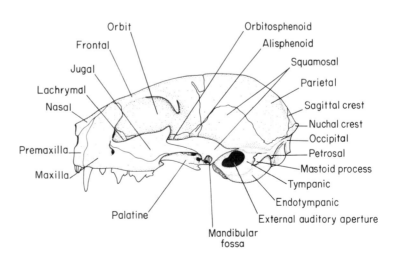

Fig. 9.4 Lateral view of a cat's skull.

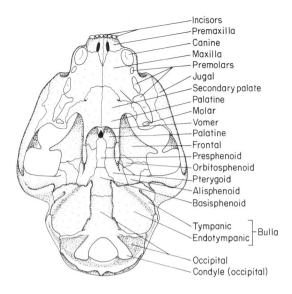

Fig. 9.5 Palatine view of a cat's skull.

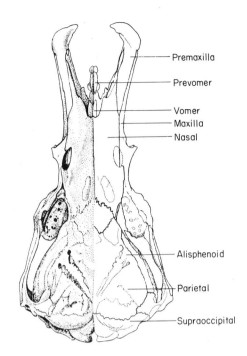

Fig. 9.6 The skull of *Ornithorhynchus*, the duck-billed platypus, seen from above.

auditory region and contain the sella turcica in which the hypophysis is lodged. An alisphenoid occurs at the side of the basisphenoid, and presphenoids form the most anterior part of the floor of the braincase. On either side a plate-like orbitosphenoid is continuous with the presphenoid. Further forward there is, in many mammals, a mesethmoid component. Scrolls of bone derived from this, and the inner parts of the nasals and maxillae, subdivide the nasal cavity. Such turbinal or turbinate bones are particularly well developed in the goat antelopes—musk ox, chamois, goural, takin, serow, saiga, etc.—which live in cold habitats. It is assumed they ensure that the inhaled air is warm prior to entering the pulmonary region.

iv Particular interest attaches to those characters which distinguish the skulls of the three principal taxa (Figs 9.6 and 9.7). The skulls of the monotremes are quite characteristic. In front of the nasals and maxillae the flattened premaxillae of *Ornithorhynchus* form the os carunculae which gives the duck-bill appearance. A jugal is absent from echidnas and the bones of the skull are so strongly ossified that they tend to fuse together. The braincase is limited to the occipital and parietal regions and the nasalia, maxillae and premaxillae are prolonged

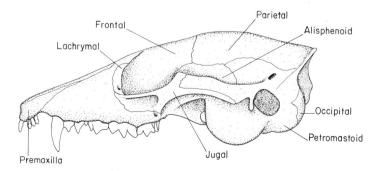

Fig. 9.7 Lateral view of a marsupial skull.

134   *The Mammals*

into the long snout. This prominent olfactory region is associated with seven vertical endoturbinal and numerous ectoturbinal bones that are referred to collectively as ethmoturbinals. In addition there are nasoturbinals and maxilloturbinals.

v A perforated bony palate is common to both the marsupials and placentals such as the Insectivora and rodents. However, a marsupial skull can always be distinguished from those of placentals by the following combination of characters:

> an inflected lamina on the internal surface of the lower jaw;
> a thickened hind-end to the palate;
> the lachrymal bone extending out of the orbit;
> a bulla, if present, of alisphenoid origin;
> the jugal running far back across the zygomatic arch.

(b) The vertebral column

i The axial skeleton is more differentiated than is the case in reptiles. With three exceptions there are always 7 cervical vertebrae. The exceptions are the sloths, *Bradypus* and *Choloepus*, with 9 and 6 or 7 respectively, and the Sirenia with 6. The first two cervical vertebrae typically comprise the atlas and axis (Fig. 9.8). The atlas consists of a ring representing its neural arch and hypocentrum, whilst the axis has the centrum of the atlas attached to its anterior end as the odontoid process. Together they permit extensive rotation. An idiosyncratic condition occurs in some Cetacea and is epit-omized by the situation in phocaenids where the first 6 cervical vertebrae are fused into a single unit.

ii In more posterior regions 5–7 ribless lumbar vertebrae contrast with the 12–14 rib-bearing thoracics. In the 3–13 sacral vertebrae the originally separate ribs are lost and replaced by transverse processes.

(c) The girdles and sternum

i The pectoral girdle of monotremes is very reptilian in appearance but in most taxa it is modified by the reduction or disappearance of certain bones. The sternum is always a conspicuous skeletal feature and may be broadly subdivided into an anterior presternum, a mesosternum with sternebrae, and a posterior xiphisternum. All the longer dorsal ribs are attached firmly to it and the whole thoracic structure is an integral part of the pulmonary ventilation mechanism.

ii In *Tachyglossus* the pectoral girdle consists of 2 scapulas, 2 coracoids, 2 clavicles, 2 precoracoids (epicoracoids) and a median interclavicle (Fig. 9.9). The dorsal scapula together with the ventral coracoid and precoracoid form a unit on each side which is attached to the rib-cage by the scapula and to the sternum by the united coracoid and precoracoid. The glenoid cavity is a wedge-shaped excavation at the point of scapulocoracoid union. The paired clavicles lie along the cross-member of the T-shaped interclavicle, united in turn to the sternum behind, while the T is clamped to the scapulas anteriorly. The triple binding so provided accounts for the echidna's powers of digging.

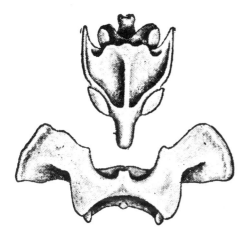

Fig. 9.8 Atlas and axis vertebrae of a hare.

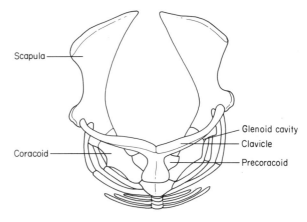

Fig. 9.9 The pectoral girdle and ribs of *Tachyglossus*, an echidna, seen from the front.

iii The interclavicle, precoracoid and coracoid are absent as distinct entities from the pectoral girdles of most mammals. The scapula bears a prominent ridge on its outer surface which projects as the acromion process, that is connected to the sternum by the clavicle. This last overlies, and unites with, the embryonic precoracoid. When present the compound structure rests on the anterior end of the sternum but it is often absent, particularly in cursorial forms where the anterior limbs swing in the line of the long axis of the body. Thus it is much reduced in dogs and absent from horses, Cetacea and Sirenia.

The Insectivora retain considerable vestiges of the coracoids and the coracoid processes are unusually large in Xenarthra such as the sloths and anteaters. Indeed the bones persist as separate entities in *Myrmecophaga jubata*. Vestiges of the precoracoid occur at the end of the primate and rodent clavicles.

iv In all mammals apart from the Sirenia and Cetacea the pelvic girdle is fused to the lateral parts of the sacrum and comprises two innominate bones each including an ilium, ischium and pubis (Fig. 9.10). The latter two meet their contralateral homologues in the midventral line at the ischiopubic symphysis, and the ischium is separated from the pubis by the obturator foramen. In echidnas and marsupials two epipubic bones are attached to the pubis and support the pouch or incubatorium. The ischiopubic symphysis is weak in Insectivora and restricted to the pubic region in *Erinaceus*. It is short in Xenarthra and, as in Chiroptera, the pelvis is here united to the vertebral column by both ischia and ilia. In contrast the symphysis is extensive in Fissipeda, al-though either short or absent in Pinnipedia. In Cetacea and Sirenia the entire pelvis is reduced to two slender bones.

(d) The limbs

i The limbs lie parallel to the body and almost underneath the trunk. The humerus and femur typically move in a fore and aft plane and function as a lever with the fulcrum at one end (elbow or knee), the weight to be raised near the other end at the articulation with the girdle, and the muscles attached just beyond at the greater tuberosity of the humerus and the greater trochanter of the femur. The predominantly dorsal position of the principal musculature is associated with the reduction of the ventral parts of the girdles.

ii Adaptations to particular types of locomotion occur in parallel in different orders (Fig. 9.11). In very heavy mammals the 'angles' between the bones are obliterated and the whole limb becomes a vertical chain of bones forming a functional strut. The plantigrade gait of ancestral forms, and certain living species such as the hedgehog or bear, in which the entire plantar surfaces are in contact with the ground, is associated with relatively slow locomotion. The most marked adaptations for cursorial locomotion occur in the feet of species whose normal environment includes relatively hard and level ground. In this case lateral disturbances to the feet are at a minimum and the main work falls on the centre digits. Typical cursorial limbs hence show a

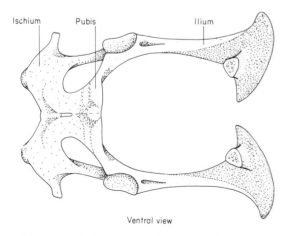

Fig. 9.10 Ventral view of the pelvic girdle of a sheep.

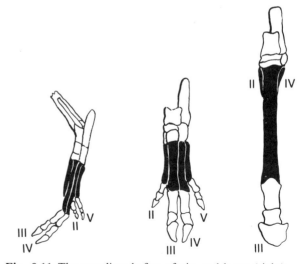

Fig. 9.11 The unguligrade feet of pig, and horse (right). (After Villee *et al.*)

reduction in the size of, or a total loss of, the lateral digits—e.g. the Equidae, cows, and fossil forms such as the Litopterna. There is also an increased limb length which is achieved by lifting the carpals and metacarpals, tarsals and metatarsals, up from the ground and adding them to the main shaft of the limb. At the same time the metapodial components are elongated. The limb thus becomes digitigrade or unguligrade. Other quite different specializations characterize the Chiroptera, with their elongated forelimbs supporting the wing membrane, and the Cetacea in which the limbs form flippers.

iii If the centre of gravity lies approximately halfway between the hip and shoulder the fore- and hind-limbs are of approximately similar size. The alternative condition, noted under dinosaurs, also occurs in mammals. If owing to a large tail, or other cause, the centre of gravity lies nearer to the hip, then more weight rests on the hindfeet than on the forefeet and the hind legs are much more powerful than the forelegs. A sudden powerful contraction of the retractor muscles of the hind legs is then capable of rotating the whole front of the body upwards. Kangaroos and many rodents also give a simultaneous thrust with the front legs which projects the front end of the body backwards until the

centre of gravity lies over the hind feet. Relatively little muscular effort is then needed to remain in this position, apart from that required to prevent any loss of balance, and such factors underlie the bipedalism of primates, kangaroos, etc.

THE MUSCULAR SYSTEM

i The skeletal musculature contributes between a third and a half of the body weight. Most individual muscles are attached to a bone or cartilage by connective tissue which can take the form of a cordlike tendon or, alternatively, comprise a sheetlike aponeurosis. However, some muscles are attached directly to the periosteum of a bone.

ii The embryonic development of the head musculature facilitates grouping the individual components (Fig. 9.12) into six associations—facial, masticatory, tongue, pharyngeal, laryngeal and ocular. The superficial components of the facial association, which play such an important part in communication, are derived during development from three primary layers of the primitive sphincter colli. The masticatory muscles vary considerably depending upon whether the animal is a herbivore, omnivore or carnivore (Fig. 9.13). The

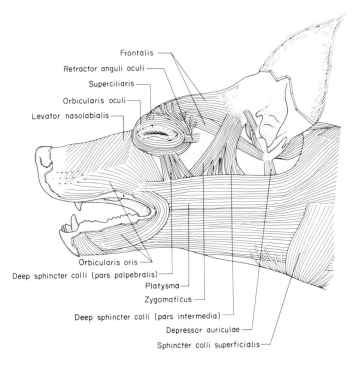

Frontalis
Retractor anguli oculi
Superciliaris
Orbicularis oculi
Levator nasolabialis

Orbicularis oris
Deep sphincter colli (pars palpebralis)
Platysma
Zygomaticus
Deep sphincter colli (pars intermedia)
Depressor auriculae
Sphincter colli superficialis

Fig. 9.12 The facial muscles of a dog. (After Miller *et al.*)

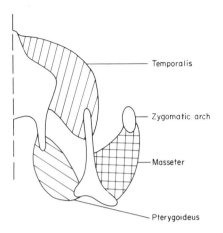

Fig. 9.13 Stylized cross-section through half the skull of a fossil triconodont showing the muscles associated with jaw movement.

masseter is particularly developed in herbivores, the temporalis is the most prominent in carnivores, whilst pterygoideus and digastric elements also contribute to the lateral displacements of the jaw in herbivores.

iii The trunk muscles vary somewhat in different taxa but are conveniently grouped into those associated with the vertebrae; those contributing to the lateral and ventral thoracic wall and including the intercostals; those of the abdominal wall, and those associated with the tail. The special muscles of the trunk are partially overlain by those passing from the trunk to the limbs.

The first of the groupings (Fig. 9.14) includes the trapezius aggregation which, crossing the interscapular

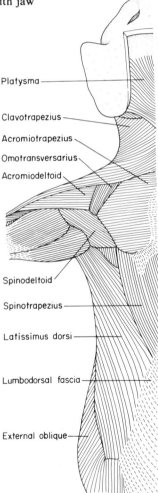

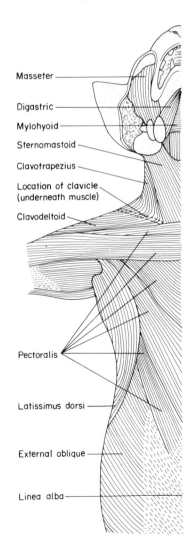

Fig. 9.14 Dorsal (left) and ventral (right) views of the musculature of a cat. (After Holmes.)

138   *The Mammals*

region of the shoulder, elevates the limb and draws it forward. The serratus, arising from a broad aponeurosis and inserted on the proximal area of the ribs, draws the last three or four ribs caudally during expiration, and the splenius contributes to raising and extending the head and neck. The longissimus system is always complex and various components extend the vertebral column or raise the anterior parts of the body.

Muscles of the abdominal wall such as the external and internal obliques compress and support the abdominal viscera. The rectus abdominis, a long, wide, flat muscle, is involved in all those functions that necessitate an abdominal pressure—expiration, urination, defaecation, parturition, supporting the abdominal viscera, drawing the pelvis forward or flexing the back. The transitory pouch of female echidnas is formed by thin skin but, in contrast, that of marsupials is often permanent and associated with a definitive sphincter muscle derived from the postero-ventral pudendal part of the panniculus carnosus muscle (Fig. 9.15). Behind these a large number of small muscles enclose the

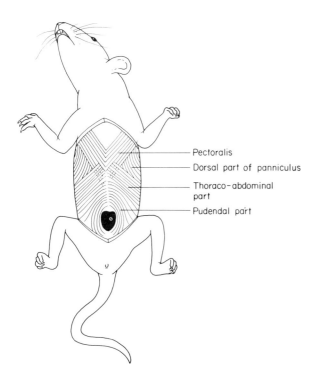

Fig. 9.15 Diagram showing the pudendal part of the panniculus carnosus muscle that forms the musculature of the marsupial pouch. (After Enders.)

Labels in figure:
Pectoralis
Dorsal part of panniculus
Thoraco-abdominal part
Pudendal part

coccygeal vertebrae and are responsible for raising, lowering, extending or curling the tail.

iv The muscles of the forelimbs include the trapezius (q.v.) alongside the omotransversarius, which also draws the limb forward; the rhomboideus which elevates it; the superficial pectoral which draws the limb inward, forward or backward depending upon its position, and the latissimus dorsi which, besides drawing the trunk forwards, draws the limb towards the trunk. The biceps and triceps brachii together with the brachialis flex and extend the elbow. The coraco-brachialis extends and adducts the shoulder joint; the brachioradialis rotates the radius. The extensor digitorum, of variable form according to the digital organization, extends the digits, whilst flexor and extensor carpi units flex or extend the carpal joints.

v In the pelvic and hindlimb regions various gluteal derivatives extend and abduct the hip joint whilst the obturator and gemelli rotate it outwards. The biceps femoris extends the hip, flexes the knee and levates the tibia.

THE DIGESTIVE SYSTEM

(a) The buccal cavity

i The buccal cavity is bounded by the lips, contains the tongue, teeth, hard and soft palates, and receives the secretions of salivary glands. The lips comprise a thin, stratified, squamous keratinized epithelium. Hairs, together with sebaceous and sweat glands, increase in number at greater distances away from the edges of the lips. In large ruminants the upper lip is modified towards the midline to form the muzzle. This consists of hexagonal areas separated by shallow trenches. In small ruminants, such as the sheep, and in pigs the upper lip unites with the nose.

ii Monotremata only have transitory ill-formed teeth. Mammals generally have milk teeth in the young that are replaced by the permanent dentition later. The permanent tooth rows of generalized insectivorous forms are long and almost parallel. Incisors are visible in side view and are sharp; canines are larger than incisors; molars and premolars are sharp-cusped and short-crowned, e.g. Insectivora, Chiroptera, Didelphidae and Phalangeridae. In the omnivorous genera the palate retains a similar shape but the incisors are disposed more transversely. The molars are bunodont and

Table 9.1 Some common descriptive terms applied to the teeth of mammals.

Term	Meaning
Bunodont	Low rounded cusps
Bunolophodont	Intermediate mixed condition
Lophodont	Cusps fused into ridges
Hypsodont	High crowned
Selenodont	Cusps expanded into crescents

brachydont, e.g. Suiformes and monkeys. In herbivores there are either cutting incisors or alternatively horny pads, as in the Bovidae. The canines are usually absent and a gap, the diastema, separates the incisors and cheek teeth. The premolars and molars form a series of grinding teeth. They are broad-crowned, hypsodont, and exhibit a tendency to produce a complicated crown pattern. Examples abound in the Multituberculata, the kangaroos, Lagomorpha, Rodentia, Artiodactyla and Perissodactyla (Fig. 9.16).

In carnivores the palate is short, and wide behind, the incisors are for gripping, the canines large, and one pair of teeth may form shearing carnassials whilst other molars and premolars can be reduced or absent.

iii The hard palate is particularly thick in ruminants and its rostral portion forms the dental pad. The tongue can bear both mechanical and gustatory papillae. The first-named type includes the filiform papillae (seen in cats and ruminants and used in the actual prehension of food) and the lenticular papillae which prevent food from slipping off the tongue. Gustatory papillae include

fungiform, circumvallate and foliate structures. Taste buds, with their constituent sensory and supporting cells, are present, in particular, on the medial, and sometimes the lateral, wall of the trench surrounding circumvallate papillae. In neonates, temporary cone or leaflike papillae are distributed around the edge of the tongue and help to produce a vacuum during suckling.

iv Salivary glands are packets of glandular tissue classified as mucous, serous or mixed on the basis of their secretions. Amongst the serous grouping, which produces amylase, are the ventral buccal glands of dog, horse and sheep, and the labial glands of the horse, pig and larger ruminants. The mucous group, which produce saliva, includes the labial and lingual glands of sheep, goats and carnivores.

(b) The stomach

i The buccal cavity is separated from the cardiac sphincter of the stomach by the pharynx and oesophagus. A true glandular stomach with gastric pits (Fig. 9.17) is present in carnivores and a non-glandular region is absent. A small initial non-glandular region occurs in the Suiformes and Equidae, whilst all the fore-stomachs of the ruminants are non-glandular. Here the rumen, a large reservoir for food in which cellulose is broken down by bacteria and Protozoa, is characterized

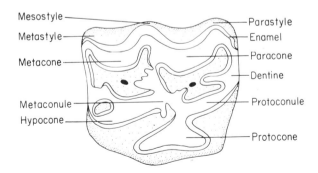

Fig. 9.16 Right upper molar of a horse showing the lophodont (ridged) structure and various other components.

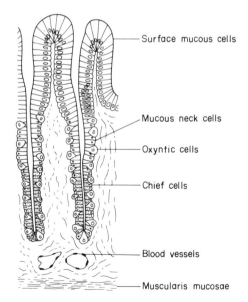

Fig. 9.17 Stylized diagram of a gastric pit and the component cells.

by wide, flat, ruminal papillae which contribute to heating and absorbing the ruminal contents. The reticulum is separated from the rumen by a fold and its mucosa is thrown into primary, secondary and tertiary folds, as is that of the omasum. The abomasum and simple stomachs have numerous gastric pits.

ii An interesting parallel to the ruminant stomach is provided by that of kangaroos (Fig. 9.18). The oesophagus opens into a funnel-shaped region which extends along the inner curvature as a partly closed, spiral groove lined by non-glandular, stratified squamous epithelium. This is analogous to the oesophageal groove of ruminants which directs non-fibrous food past the reticulo-rumen to the abomasum. In macropods the spiral groove communicates with a non-sacculated, thick-walled chamber which, with the pyloric region, serves similar functions to those of the ruminant abomasum. It has a full complement of parietal and chief or pepsinogenic cells, and a highly acid pH of 1.8–3.0. The larger, sacculated region of the stomach bears cardiac glands that secrete mucus but not proteolytic enzymes or acid. The pH ranges from 4.6 prior to feeding to 8.0 afterwards. During pouch-life all regions of the neonate's stomach exhibit proteolytic activity and a variable but acid pH.

The sacculated region has a similar function to that of the reticulo-rumen of ruminants and also contains a dense population of cellulytic bacteria (10^{10} cm^{-3}) and protozoans (10^6 cm^{-3}). In both cases the fermentation chamber produces volatile fatty acids—ethanoic, propanoic and butanoic acids—at a rate of 20 μmol g^{-1} h^{-1}.

In macropods their concentration is very low in the fundic region of the stomach and they are presumably absorbed into the portal blood system in the cardiac region.

(c) The intestines

The small intestine links the stomach, of whatever form, to the large intestine. Three sections are typically identified—the duodenum, jejunum and ileum. All three possess villi, and crypts of Lieberkuhn can be distinguished histologically. The duodenum has high and sometimes leaflike villi, together with submucosal glands, and receives the pancreatic and bile ducts. The jejunum has fewer smaller villi, and the ileum has lymphoid (Peyer's) patches and some deep craters, especially around the ileocaecal region. In those herbivores which have a simple stomach (e.g. Lagomorpha and Equidae) the caecum is an important reservoir within which bacterial fermentation takes place, and Lagomorpha increase its effect by passing their food through the alimentary canal twice as they eat the mucoid faecal pellets resulting from the first passage.

The large intestine always lacks villi and there is a gradual increase in the prominence of mucus-secreting goblet cells distally.

(d) The pancreas

The exocrine portions of the pancreas comprise a large tubulo-acinar gland divided into lobules by connective tissue trabeculae. The acini comprise 2–15 pyramidal

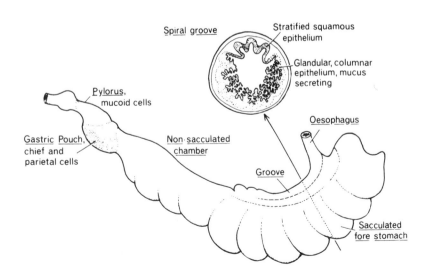

Fig. 9.18 A kangaroo's stomach uncoiled to display the regions. A section of the sacculated region is shown above.

cells whose nuclei are surrounded by rough endoplasmic reticulum and numerous polyribosomes. There is a well-developed golgi apparatus and membrane-bound zymogen granules are located in the apical cytoplasm. A few microvilli are present apically and desmosomes link the cells together. As food enters the pyloric antrum and duodenum the hormones secretin and pancreozymin are released into the bloodstream. Secretin induces the release of a large volume of pancreatic fluid with a high Na_2CO_3 content that neutralizes the stomach acid in the food bolus. Pancreozymin enhances zymogen secretion.

(e) The liver

The liver is the largest gland of the digestive system and receives, via the portal vein, most of the substances absorbed from the small intestine. The connective tissue is generally thin but is thick in Artiodactyla, where connective tissue also delineates the individual lobules. These have afferent interlobular vessels, sinusoids, lobule cells and the centrilobular collecting vein that drains into the posterior vena cava. The overall organization, therefore, gives rise to a functional gradient with the more peripheral lobular regions receiving oxygen and nutrients before more centrally situated ones. A parallel differential cell necrosis also characterizes various pathological conditions and enables post-mortem differentiation of death from alcohol poisoning or typhoid, for example.

RESPIRATORY SYSTEM

The respiratory system is more complex than in reptiles. It consists of an air-conducting portion from the nose through to the tertiary bronchioles, and a gaseous exchange portion including the respiratory bronchioles and the alveoli. The walls of the hollow organs comprise mucosal, submucosal, muscular and serous tunics. The vestibule of the nasal cavity is often lined by stratified non-keratinized epithelium but may be lined by skin. The respiratory nasopharynx lies above the soft palate and the oropharynx is common to both the digestive and respiratory systems as it lies behind and below this. The lung consists of lobes and lobules and within these the respiratory bronchioles give origin to alveolar ducts from which many alveoli open. These are polyhedral air spaces closely related to blood vessels and lined by a simple epithelium. Air exchange within the system is under the control of the Hering-Breuer reflex and respiratory reflexogenic zones of the pons and medulla oblongata.

URINOGENITAL SYSTEMS

(a) The urinary components

In general terms the kidney resembles that of other amniotes. It has a radial organization with an external capsule surrounding the cortex and one or more medullary pyramids. Each nephron is divisible into a renal capsule composed of glomerulus and Bowman's capsule; a proximal tubule with successive convoluted and straight portions; an intermediate tubule with descending and ascending limbs, and a distal tubule with straight and convoluted portions. The secretory and absorptive functions of these components are summarized in Fig. 9.19. The ureter leads to the urinary bladder which is evacuated by the urethra. In monotremes and marsupials this opens into the cloaca via a

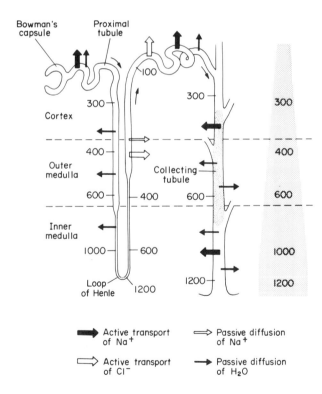

Fig. 9.19 Summary of the fluxes along a mammalian renal tubule. The shaded portion of the collecting duct is ADH sensitive. The osmotic gradient in the extracellular fluid is shown on the right (mosmol l^{-1}).

distinct urinary papilla that is distinct from the genital papilla of the male penis, but in placentals it receives the male genital ducts and opens at the tip of the penis.

(b) The male genital system

i The testis comprises a tunica vaginalis, a tunica albuginea and seminiferous tubules. Spermatogenesis takes place under the control of hypophyseal FSH whilst LH stimulates the associated cells of Leydig. Two types of spermatogonia can be distinguished. The type A cells lie close to the basement epithelium and have finely dispersed chromatin plus one or two irregular nucleoli. Type B cells occupy a more central position and the chromatin is often clumped towards the nuclear envelope. The Type A cells divide mitotically. Type B cells presumably originate from Type A cells and give rise to primary spermatocytes by further mitosis. These then undergo meiosis to give spermatids. The actual spermiogenesis can be subdivided into a Golgi phase, during which proacrosomal granules appear in the Golgi apparatus; a cap phase during which there is a progressive increase in the adherence surface between vesicle and nucleus, giving the head cap; an acrosomal phase, during which much of the acrosome forms the acrosomal cap, and, finally, a maturation phase during which the final spermatozoan appearance is attained (Fig. 9.20).

ii Among many marsupials and all placentals the optimum temperature for spermiogenesis is several degrees lower than that of the body. This has led to adaptations such as the external scrotal sac which is anterior to the penis in marsupials, posterior to it in placentals. If the ambient temperature falls too low the testes are drawn nearer to the body by reflex contraction of the cremasteric muscle.

In placentals the spermatic artery emerges from the inguinal canal and is then coiled and surrounded by the branched spermatic vein giving the pampiniform plexus. The coiling of the artery damps the pulse pressure but leaves the mean blood pressure and oxygen concentration unaffected. Heat exchange between the arterial and venous blood then reduces the temperature by some 5°C relative to the femoral artery. An analogous but anatomically distinct counter-current heat exchange occurs in some marsupials where the spermatic artery divides to produce an elaborate rete mirabile. This is absent from *Notoryctes*, the marsupial mole, whose testes are in the body wall, and from small genera such as *Sminthopsis* and *Acrobates*. It is best developed in

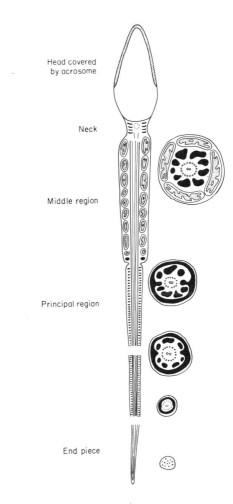

Fig. 9.20 Diagram of a mammalian sperm.

species with a large body such as the Tammar, with 154 arterial branches, and the Tasmanian devil with 15. In addition, many marsupials have a deeply pigmented tunica vaginalis around the testes which presumably acts as a black box radiator in high ambient temperatures.

iii The epididymis consists of efferent ductules which connect the testis and vas deferens. The number of ductules varies from 13–19 in common placentals and they are lined by a simple epithelium with occasional cilia. Microvilli occur on the luminal surfaces and isolated or grouped non-ciliated cells have secretory characteristics.

iv Accessory glands vary. Placentals such as the

The Mammals 143

stallion, bull and boar have vesicular glands but these are absent from common carnivores. Both marsupials and placentals have a prostate gland. Its greater complexity in marsupials has been explained as reflecting the incorporation of other accessory glands into a single structure. In all cases it contributes to the nutrition of the sperm and has a high phosphatase and fructose content. Bulbo-urethral glands are widespread. The volume of the resulting ejaculate is relatively greater, though the sperm density is lower, in the kangaroo than in placentals such as the bull or ram.

v The penis includes vascular erectile tissue—corpora spongiosa and cavernosa—which when engorged gives an erection. In common placentals two types of penis occur. The vascular type is epitomized by the Equidae, has little connective tissue, and contrasts with the fibrous type in Artiodactyla and Carnivora. Erection begins with the relaxation of smooth muscle fibres in the helicine arteries that open into the caverns of the erectile tissue. The main blood flow is then directed into these caverns and compression of the smaller veins decreases outflow from both the caverns, and, where it exists, the glans penis. In some species, such as the dog, the constrictor vulvae muscle of the bitch enhances this effect, engorgement prevents immediate withdrawal, and coitus is prolonged. Detumescence is initiated by contraction of the muscles in the helicine arteries. Contraction of the smooth muscle and elastin fibres in both the tunica albuginea and the related connective tissue trabeculae then returns the penis to a flaccid state.

(c) The female genital system

i The female of a viviparous species has to receive the male ejaculate and also have a tract able to secrete the egg coats and nourish the embryo. Some mammals have a long period of oestrous, when the tract is receptive to spermatozoa, followed, either spontaneously or in response to copulation, by a progestational phase. Another pattern involves cyclical recurrence of oestral and progestational phases. A monoestrous pattern occurs in some dasyurids amongst marsupials (q.v.). However, the majority of marsupials and placentals exhibit polyoestrous although no marsupial is known to have a cycle as short as those of murid rodents.

ii The ovary, surrounded by a tunica albuginea, usually consists of a peripheral cortex and a central vascular zone but in mares this situation is reversed. Embedded in the stroma of the cortex are follicles in various states of development. Primordial follicles contain an oocyte, develop under the influence of FSH, and move to the deep cortical layers. The oocyte, surrounded by cells in a radiate arrangement—the corona radiata—becomes located in a thickening of the granulosa components of the follicle. This is the cumulus oophorus. Mature follicles bulge from the ovary walls and the follicular wall thins just prior to ovulation. It is at this time that the first maturation division occurs giving a secondary oocyte and first polar body. The second maturation division follows but remains at metaphase until fertilization when the second polar body is formed. Meanwhile, after ovulation, the follicular wall collapses. Under the influence of LH both the granulosa and theca interna cells then proliferate giving a corpus luteum with granulosa lutein cells centrally and theca lutein cells peripherally. If fertilization does not occur this corpus luteum is transformed to a corpus albicans.

iii The fallopian tubes vary in form but their lining epithelium always contributes nutrients, and ciliated cells move the ovum. The uterus, cervix and vagina represent the fused portion of the Mullerian ducts and the differing overall structure in marsupials and placentals is represented in Fig. 9.21. In all cases the developing blastocyst embeds in hypertrophied uterine epithelium. However, in marsupials the sperm reach the ovum via the lateral vaginae whilst the young are born via a central birth canal. This forms through the connective tissue strand that connects the median vaginal chamber to the posterior wall of the urinogenital sinus. In kangaroos this birth canal then remains patent throughout subsequent life but in most marsupials it is repaired by scar tissue.

iv The placentae which nourish the young and serve for gaseous exchange and excretion are of varying structure and origin. In echidnas the endogenous yolk only suffices to sustain development to a bilaminar vesicle and subsequent development relies upon uterine secretions. In both these animals and marsupials only part of the yolk-sac is vascularized. The echidna embryo has c. 19 somites when the egg is laid and development continues up to hatching in the maternal pouch without additional nourishment so that during this time the allantois is only involved in gaseous exchange.

In marsupials the outer surface of the yolk-sac wall develops a weft of microvilli with pinocytotic vesicles at their bases. Mitochondria and rough endoplasmic reticulum are both widespread. In the Tammar the

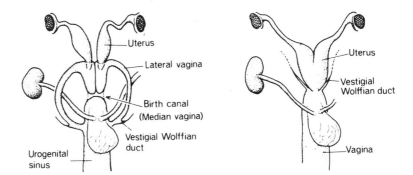

Fig. 9.21 A comparison between the female systems of a marsupial (left) and placental (right). (After Fordham, from Tyndale-Biscoe.)

microvilli interdigitate and, as the shell membrane disintegrates, this surface comes into contact with the uterine eipthelium. In *Philander, Dasyurus* and *Phascolarctos* there is a more intimate association between the two which increases in later stages of pregnancy, and vitelline cells come into close association with the maternal circulation. In *Didelphis* the allantois reaches its maximal development 12–24 hours prior to birth. The most elaborate development of this yolk-sac or allantoic placenta occurs in *Perameles*. Here the ectoderm of the chorion gives rise to a superficial syncytium which erodes the adjacent uterine epithelium and forms an intimate structure in which the fetal and maternal capillaries lie close together. The young of this genus are more developed at birth than is the case in other marsupials. Much discussion has centred on whether this chorio-allantoic placenta reflects parallel evolution with the placental condition or whether both derive from the structure in a common ancestor.

The placentae of Eutheria have been variously classified on the basis of their shape and distribution; the layers intervening between fetal and maternal circulations, or the depth and extent of implantation. Differentiation on the second basis is represented in Table 9.2.

ENDOCRINE ORGANS

(a) The hypophysis

As in other vertebrates the hypophysis has close functional relationships with the overlying hypothalamus and despite an underlying uniformity many specific variations in the structure exist. The overall form varies from the compressed and elongated organ of horses to the flattened organ of pigs and dogs. The adenohypophysis has at least seven functionally different cell types responsible for secreting prolactin, somatotrophic

Table 9.2 Placental characteristics based upon the degree of apposition between maternal and fetal blood.

Placenta type	Characteristics	Examples
Epitheliochorial	Chorionic and uterine epithelia remain intact	Equidae, Suidae, large ruminants
Syndesmochorial	Histolytic action of the chorionic epithelium erodes the uterine epithelium	Deep parts of placentoma in small ruminants
Endotheliochorial	Further histolysis by chorionic epithelium erodes uterine connective tissue	Most zonal placentae of carnivores
Haemochorial	Direct contact between chorionic epithelium and maternal blood	Some rodents and primates and the lateral parts of carnivore placentae

hormone, adrenocorticotrophic hormone, follicle stimulating hormone, thyrotrophic hormone, luteinizing hormone and MSH respectively. Release of the hormones into the blood stream is under the control of releasing factors from the hypothalamus.

(b) The thyroid gland

The capsule and trabeculae of the thyroid are often thin and the main constituent structures are follicles consisting of a centrally located colloid with a surrounding epithelium. The extracellular storage of the hormone is a unique characteristic, but secretory droplets of varying size occur in association with the Golgi apparatus and are probably under the control of lysosomes. Parafollicular cells in the follicular epithelium, or in interfollicular positions, are the source of calcitonin.

(c) The parathyroid glands

These glands develop from pharyngeal pouches III and IV. Those derived from III are closely apposed to the thymus, those from IV adjacent to the thyroid. The component cells are arranged in clusters or strands, and comprise light chief cells; active dark chief cells; oxyphilic cells and syncytial cells. Parathormone maintains the blood calcium at normal levels by promoting withdrawal of calcium from bone; by indirectly increasing intestinal absorption by initiating the release of 1, 25 dihydroxycholecalciferol by the kidney, and by decreasing urinary loss. In parathyroid deficiency the fall in serum calcium and the rise in inorganic phosphorus leads to tetany. Hyperactivity gives low serum phosphorus, high urinary phosphorus and very high serum and urinary calcium levels.

(d) The adrenal glands

i The mesodermal and neural crest derivatives are fused into single anatomical structures in mammals, the former forming the distinct cortex and the latter the medulla. The cortex is divisible into four zones whose steroid biogenesis is either under the control of hypophysial adrenocorticotrophic hormone or, as in the case of the zona glomerulosa, stimulated by angiotensin II:

the zona glomerulosa with the constituent steroid secreting cells in cords;

the zona intermedia separating the glomerulosa and fasciculata;

the zona fasciculata with radially arranged cords of cells each in contact with a sinusoid;

the zona reticularis, a continuation of the fasciculata and comprising an irregular network of anastomosing cell cords.

ii The zona glomerulosa produces mineralocorticoids, deoxycorticosterone and aldosterone. These maintain the electrolyte levels of the extracellular fluids by controlling reabsorption from the renal tubules. Following adrenalectomy there is an increased release of sodium and an increased retention of potassium. Low sodium levels lead to a drop in blood pressure that stimulates the juxtaglomerular cells of the kidney to secrete renin. This leads to the conversion of angiotensin

I to angiotensin II which stimulates aldosterone secretion and induces vasoconstriction.

iii The zona fasciculata and the zona reticularis produce the glucocorticoids cortisone, hydrocortisone and corticosterone. Hydrocortisone facilitates protein catabolism and gluconeogenesis. Other corticosteroid functions involve the destruction of lymphocytes and release of γ-globulins, the decrease of circulating eosinophils and anti-inflammatory effects. Both oestrogens and androgens are also produced and if the latter are excessive they induce masculinization of females.

iv In the medulla, noradrenalin-producing cells possess large nuclei and the cytoplasm contains argentaffin granules. Adrenalin-producing cells are similar but have a high affinity for azocarmine.

(e) The islets of Langerhans

The endocrine components of the pancreas are surrounded by reticulin fibres and include five types of cells (A–F) in irregularly anastomosing cords. The exact percentage representation of each type varies differentially with age and species. Insulin is produced by B cells, glucagon by A cells (see page 116).

CARDIOVASCULAR SYSTEM

(a) The heart and arterial system

i The *four*-chambered heart of mammals is a compact structure lying in the mediastinal space ventromedial to the lungs. An external coronary sulcus marks the point of junction between the atria (auricles) and ventricles. The common pulmonary artery, arising from the right ventricle, divides to give right and left pulmonary arteries. The systemic aorta arises from the left ventricle, bends sharply left a short distance from its origin, and then runs in a postero-dorsal direction beneath the vertebral column. The general features are illustrated in Fig. 9.22.

ii Two coronary arteries arise near its base and supply blood to the heart itself. The next vessel which it typically gives off is the innominate or brachiocephalic artery from which the common carotids arise. Subclavians take blood to the arm and once outside the body wall are referred to as the axillary arteries. Three kinds of vessels arise from the abdominal aorta—paired parietal vessels to the body wall; visceral branches to the

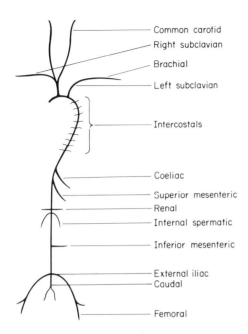

Common carotid
Right subclavian
Brachial
Left subclavian

Intercostals

Coeliac
Superior mesenteric
Renal
Internal spermatic
Inferior mesenteric

External iliac
Caudal

Femoral

Fig. 9.22 A schematic representation of the arterial system of a mammal.

abdominal organs, three of which are unpaired; and paired terminal branches supplying the hind limbs. The parietal vessels include adrenolumbar arteries, phrenic arteries, iliolumbar and lumbar arteries. The paired visceral vessels include renal and genital arteries, the latter comprising the spermatic arteries of males and the ovarians of females. The unpaired vessels are the coeliac and the superior and inferior mesenterics.

iii The external iliac arteries arise from the aorta near to the anterior end of the pelvic girdle and give off *deep femoral* derivatives just prior to leaving the body cavity. From these last there arise in turn vesicular, external spermatic and inferior epigastric derivatives. Internal iliacs arise from the aorta just posterior to the external iliacs and, outside the abdominal cavity, the last-named are referred to as the femorals.

(b) The venous system

i Blood from the hind limbs and the sacral regions returns to the abdominal cavity via the common iliac veins and a small caudal vein. Ovarian or spermatic veins, renal veins and adrenolumbar veins also conduct blood into the posterior (inferior) vena cava. The large and important hepatic portal vessel drains the

abdominal digestive system and is formed by the confluence of various vessels such as the pancreaticoduodenal, the coronary vein of the stomach, the gastrosplenic, gastroepiploic, inferior and superior mesenteric veins. Upon entering the liver it breaks into branches supplying the various lobes and lobules and its divisions ultimately end in the hepatic sinusoids. Blood from these sinusoids then collects in the centrilobular veins and passes to the posterior vena cava. As this penetrates the diaphragm it also receives small phrenic veins.

ii The anterior (superior) vena cava is a large vein in the anterior part of the body and forms the final common path through which blood from the neck, forelimbs and thoracic region reaches the heart. External jugulars drain the head and face of each side and the small internal jugular receives blood from the cranial cavity via the inferior cerebral vein. An azygos vein drains the dorsal thoracic wall and the dorsal part of the abdominal wall. The anterior vena cava itself is formed by the union of two innominate veins just cephalad to the azygos vein. These are formed on each side by the junction of the external jugulars and the subclavian veins which comprise a final path for blood from the subscapular and brachial veins. A prominent vertebral vein enters each innominate just in front of their union to form the anterior vena cava.

(c) The lymphatic system

i Lymph vessels also contribute to tissue drainage and begin as lymph capillaries in the connective tissue. Large lymph vessels drain into the venous system after filtration at lymph nodes, encapsulated aggregations of lymphoreticular tissue, where lymphocytes gain access to the efferent fluid.

ii The thymus and spleen are important components of this system. The first-named originates from the third pharyngeal pouch, controls the number of lymphocytes and regulates the immunological competence of the animal. The spleen produces blood cells during embryonic and fetal life, and acts as a major blood reservoir later.

SPECIAL SENSE ORGANS

Smell, vision and hearing all play significant roles in the detection of predators, prey or food, the recognition of the sexes and identification of young by their parents, or parents by their young.

(a) The olfactory epithelium

Smell can be the paramount sensory modality, as in macrosmatic genera like *Tachyglossus*, *Myrmecophaga*, *Manis*, etc., and is always important, even in so-called microsmatic forms such as the apes and humans where it plays crucial roles in feeding, in the perception of pheromones, etc. The olfactory epithelium is borne on the ethmoturbinates, parts of the dorsal turbinates, the nasal septum and the vomero-nasal or Jacobsen's organ, and the sensory units are bipolar nerve cells which extend through the entire height of the epithelium. A dendrite reaches the free surface and ends in a bulbous enlargement—the olfactory rod or vesicle. Cilia—olfactory hairs—8–10 μm long, and located on the surface, form the actual detectors. Up to 16 different categories of sensory cells have been defined on the basis of cell size and volume, axon diameter and number of neurotubules.

(b) The eyes

Vision is an important sensory modality in all mammals apart from a limited number of subterranean genera, such as the marsupial and placental moles and the rodents *Heterocephalus* or *Spalax*. The eye is, therefore, uniformly well-developed and differs from that in birds in relatively few ways. The eyeball is spherical, the retina bears blood vessels, and there is no structure comparable with the avian pecten (q.v.). The pineal organ is glandular and secretes melatonin.

(c) The ear

Most mammals have a well-developed sense of hearing. Many rodents use ultrasonic frequencies and these are of great importance to the echolocating bats which, with the exception of *Rousettus*, are all microchiroptera. The incorporation of the articular and quadrate bones into the middle ear, is, of course, of outstanding significance in paleontology as it provides one of the important criteria for establishing that a fossil is a mammal—it has only one bone, the dentary, in the lower jaw.

THE EXTERNAL EAR
The pinna consists of a central plate of elastic cartilage. Its considerable mobility in many genera reflects the action of external ear muscles and is of immense importance in accurate sound-localization. The external auditory meatus is surrounded by elastic cartilage and, towards the tympanic membrane, by bone. Its skin

lining bears apocrine sweat glands known as ceruminous glands, which secrete wax.

THE MIDDLE EAR
The tympanic membrane, consisting of several layers, separates the external auditory meatus from the middle ear (Fig. 9.23). The tympanic cavity is limited by this membrane laterally and by bone medially. The malleus is inserted into the tympanic membrane, the baseplate of the stapes is surrounded by a fibrous ligament and fits into the oval window, and the two are joined by the incus.

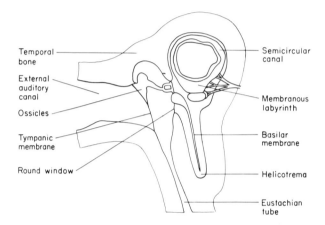

Fig. 9.23 A schematic diagram of the ear with the cochlea shown as uncoiled.

THE INNER EAR
The vestibule contains the utricle dorsally and the saccule ventrally and the semicircular canals open into it. Within the utricle and saccule there are the maculae utriculi and sacculi and the function of all these is analogous to that in birds.

The bony cochlea spirals several times around a central axis—the modiolus. An osseous lamina spirals round the modiolus and continues as the basilar membrane towards the periphery. The upper compartment of the cochlear canal is further subdivided by the vestibular or Reissner's membrane. Thus three compartments are formed—the upper scala vestibuli, the intermediate cochlear duct, and the lower scala tympani. They are connected to the middle ear by the oval (scala vestibuli) and round windows (scala tympani). The organ of Corti consists of two types of cells—supporting and sensory—and two sets of the supporting cells occur and form the inner and outer supporting arches of pillar and phalangeal cells respect-

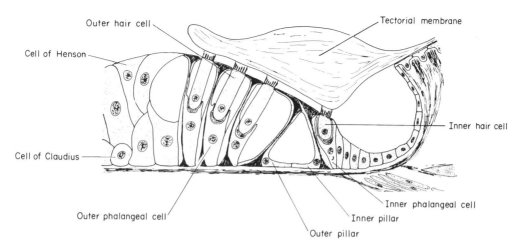

Fig. 9.24 A schematic drawing of the organ of Corti.

ively (Fig. 9.24). The basal and apical regions of the cochlea are differentially sensitive to sounds of differing frequency, and the nerve fibres conducting impulses to the cochlear nuclei of the medulla oblongata project a tonotopic representation of these sounds into the ascending auditory pathways of the brain.

CENTRAL NERVOUS SYSTEM

(a) The spinal cord

Almost all ascending and descending tracts undergo some mixing in the spinal cord. The dorsal column, consisting of the white matter lying between the dorsal median sulcus and the dorsal rootlets, is predominantly sensory and constituted by the gracile and cuneate tracts. Pyramidal corticospinal fibres also occur in some species. The lateral column, extending from the dorsal to the ventral rootlets, is a mixture of afferent and efferent tracts, whilst the ventral column, situated between the ventral rootlets and the ventral fissure, is essentially motor.

Where it exists, the spinothalamic tract conveys nociceptive impulses and spinocervicothalamic tracts include fast fibres conveying tactile impulses to the medial lemniscus. All pathways have relays in the thalamus and exhibit somatotopic localization.

(b) The medulla oblongata

As in other vertebrates the medulla contains nuclei of origin and termination of cranial nerves V – X

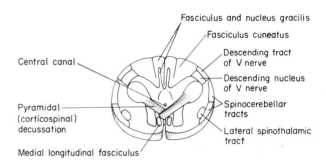

Fig. 9.25 The medullary organization of a mammal.

(Fig. 9.25). It also contains definitive olivary and pontine structures; an extensive reticular formation occupying the areas between the principal tracts and nuclei, and more or less prominent corticospinal fibres which comprise the pyramidal tracts that decussate at a variety of levels in different species.

In the anterior regions there are prominent cochlear and vestibular nuclear aggregations and both cardiac and respiratory reflexogenic zones provide autonomic control foci.

(c) The cerebellum

Comparisons between the cerebella of monotremes, marsupials and placentals remain somewhat controversial (Fig. 9.26). Amongst marsupials there is a correlation between the degree of sublobulation and the overall body-size that is reminiscent of the situation in

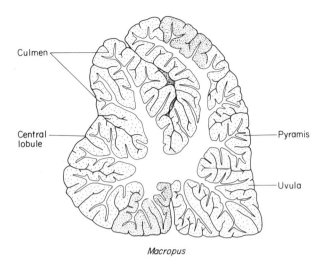

Culmen

Central
lobule

Pyramis

Uvula

Macropus

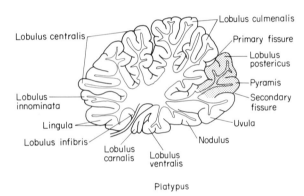

Lobulus centralis

Lobulus culmenalis

Primary fissure

Lobulus
postericus

Pyramis

Secondary
fissure

Lobulus
innominata

Lingula

Uvula

Lobulus infibris

Nodulus

Lobulus
carnalis

Lobulus
ventralis

Platypus

Fig. 9.26 Diagrammatic views of sagittal sections through the cerebella of a kangaroo (above) and a platypus (below). (After Dillon.)

birds. The relative homogeneity of the complex cerebella of placentals needs little emphasis. Problems only arise with forms such as the Cetacea in which there is a huge ventral paraflocculus. The dynamic loop hypothesis has provided a rationale for cerebellar functions. Two afferent channels provide input and, respectively, involve the so-called climbing fibres and mossy fibres. The sole efferent pathway leads, via Purkinje cell axons, to the cerebellar nuclei and brainstem reticular formation. Each Purkinje cell is supplied by a single climbing fibre but the input via mossy fibres is characterized by extensive divergence and both excitatory and inhibitory effects at the Purkinje cell. These distinct inputs have similar peripheral sources and provide a basis for theories of both feedback and feed-forward controls. In

broad terms one can envisage the cerebellar systems as providing a 'print out' of muscular activities before these have actually occurred, thereby enabling a continuous modification of motor actions as they are being performed. Cerebellar dysfunction certainly results in classic motor disturbances such as action tremors and inappropriate placing of limbs, etc.

(d) The mesencephalon

i The superior and inferior colliculi are prominent components of the mammalian midbrain. The first is the mammalian equivalent of the optic tectum in other classes and its component cells are typically binocularly driven and respond best to moving stimuli. The sharpest response peak is elicited when an *optimal* stimulus crosses a precisely defined area of the visual field. Further visual representation occurs in the pretectal nuclei.

ii The inferior colliculus is an important relay station on the ascending auditory pathways. In genera with echolocating ability its constituent cells have very precise responses to sounds of particular frequencies, usually in a high kilohertz range.

iii The red nucleus varies considerably. Distinct parvocellular and magnocellular components are absent from marsupials but are more or less distinct in placentals. The substantia nigra is also of immense importance as it is the site of formation of the dopamine that is involved in reciprocal nigro-striatal and strionigral systems which modulate the activity of the caudate/putamen complex. Dysfunction of this complex results in Parkinsonian tremor in humans.

(e) The diencephalon

i The epithalamus of mammals is not unduly dissimilar from that of birds. In the hypothalamus more or less distinct populations of cells elicit threat, mating, thermoregulatory or feeding behaviour during direct electrical stimulation. Male mounting behaviour can be elicited from females in this way.

ii The marsupial and placental thalamus consists largely of grey matter organized into an immensely complex series of nuclear aggregations which are usually associated into anterior, midline, medial, lateral and ventral arrays. Complex thalamo-cortical and cortico-thalamic projections modulate the activity within all

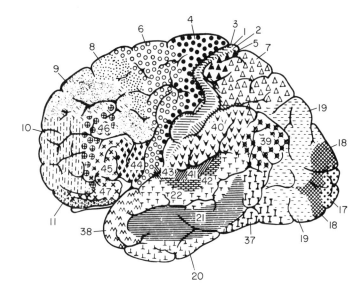

Fig. 9.27 Brodmann's cortical areas seen from the left side. Each area is designated by a number.

ascending sensory systems and there are precise somato-topic relationships between foci in the thalamic nuclei and the relevant cortical areas. The size and number of different thalamic components are broadly related to overall body size, but in aquatic or semi-aquatic forms certain nuclei are rather large by comparison with closely related terrestrial forms. The actual size of a given component also reflects the relative prominence of its sensory modality in the life of the animal. Furthermore, the nuclear aggregations of marsupials and insectivores tend to be more multimodal than are those of the remaining placental orders and this has been explained as reflecting progressive evolutionary specialization of individual nuclei.

(f) The telencephalon

The cerebral hemispheres are again amongst the most prominent features of the brain. The olfactory bulbs lie in front of, and below them. At the turn of the century Brodmann categorized the different parts of the cortex on the basis of differing cellular constitution and this provided a basis for more recent functional analyses. His cortical map of man is in Fig. 9.27 and a functional analysis of the cat occurs in Fig. 9.28.

The primary motor area (4) occupies most of the precentral gyrus and is characterized by the giant (Betz) cells in the fifth cortical layer. Stimulation of the anterior part of the precentral gyrus (4s) gives muscular

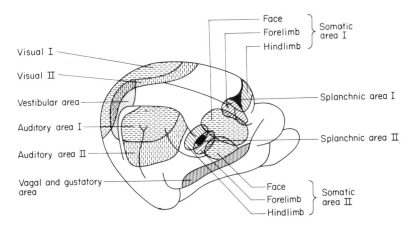

Fig. 9.28 The generalized organization of a cat's cerebral cortex as elucidated by early electrophysiological investigations.

relaxation. The cortex immediately in front of these (6) gives gross limb movements. Both the muscle relaxation and elimination of unwanted movements are achieved by the basal ganglia deep within the hemisphere. These, the caudate and globus pallidus, are involved in feedback circuits controlling the motor discharges from the cortex.

Cortex in the front (areas 9, 10, 11) constitutes association areas. Other association areas (5 and 7) lie posteriorly and their cells are active as an animal attempts to pick up, or catch, a desired object.

The primary visual area (area 17) is at the occipital pole and the more anterior areas (18 and 19) are visual association areas. The primary auditory projections are on the superior temporal gyrus (41 and 42). Deficits in the temporal regions, often resulting from strokes, lead to associative disturbances. These include both motor and sensory aphasias in which the incorrect word is spoken or perceived; alexias, inability to read; acalculias, inability to perform simple calculations, or amusias in which there is a lack of appreciation of differences of frequency or harmonic.

10 · Mammalian Diversity

MESOZOIC MAMMALS

i Fossil remains of definitive mammals occur in Jurassic deposits. Five principal orders are known—the triconodonts, symmetrodonts, docodonts, pantotheres and multituberculates. The first four of these are restricted to the Jurassic and Cretaceous; the multituberculates persist through to the early part of the Cenozoic, and the pantotheres perhaps include the ancestors of both marsupials and placentals.

ii Triconodonts and symmetrodonts have sharp teeth and were probably insectivorous. Reconstructions of the jaw musculature of triconodonts (cf. Fig. 9.13) indicate that both the temporalis and masseter muscles were reasonably prominent and this suggests that there was no great specialization towards either a carnivore or herbivore food régime.

Amongst the pantotheres the archaic family Amphitheriidae had four incisors, a canine, four premolars and seven molars. This is a larger number of teeth than occurs in any known marsupials or placentals. In contrast the family Paurodontidae had a small number.

iii The multituberculates exhibit herbivore characteristics (Fig. 10.1). There are few incisors and the lower ones are not placed vertically in the jaw. There is a diastema, and the cheek teeth show an increase in complexity through time, with those of the Jurassic genus *Plagiaulax* being less complex than those of the Cretaceous *Ptilodus*, themselves simpler than those of the Paleocene *Taeniolabis*. Reconstructions of the cranial musculature suggest that the temporalis was relatively feeble, the masseter was prominent, and the pterygoideus, responsible for contributing a lateral movement to the lower jaw in herbivores, very developed. All such characters suggest that the animals were herbivores and closely paralleled both the later lagomorphs and rodents, and the Triassic near-mammals such as tritylodonts.

Subclass Prototheria

ORDER MONOTREMATA

Many of the unique characteristics of both the duck-billed platypus and the echidnas have already been

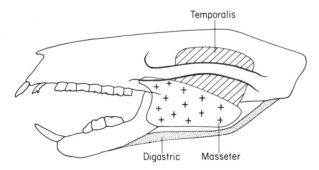

Fig. 10.1 Stylized drawing of a multituberculate skull showing the temporalis, masseter and digastric muscles; the presence of a diastema, and the front teeth at an angle.

Echidna

Platypus

Fig. 10.2 An echidna and a duck-billed platypus. (After Ride.)

referred to. Their distinctive features were long ago appreciated by de Blainville who separated them from both the marsupials and placentals on the basis of their quite separate oviducts. In broad terms one can summarize their anatomy as including definitive mammalian characters, such as hair and milk glands, together with reptilian characters such as the additional bones in the pectoral girdle. These make them quite unique. However, as is immediately obvious from a superficial inspection, they also differ quite considerably from one another (Fig. 10.2).

Subclass Theria

Infraclass Metatheria—marsupials

i The characteristic to which marsupials owe their name is not their most distinctive one. Few male marsupials have a pouch and in the females of some species it is only developed while suckling. The distinctive cranial characteristics were listed on page 135. Certain fragmentary fossils from the lower Cretaceous deposits of North America, e.g. *Pappotherium*, have been tentatively interpreted as representing conditions which are intermediate between those of marsupials and placentals. Using diverse anatomical criteria the families of living marsupials can be separated into 6 distinct groups:

Polyprotodonta
Didactyla 1 Didelphidae —American opossum
2 Dasyuridae —Native cats
Thylacinidae —Marsupial wolf
Notoryctidae —Marsupial mole
3 Caenolestidae
Syndactyla 4 Peramelidae —Bandicoots
Diprotodonta
Syndactyla 5 Tarsipedidae
6 Phascolarctidae —Koalas
Vombatidae —Wombats
Phalangeridae —Phalangers
Petauridae —Flying phalangers
Burramyidae
Macropodidae —Kangaroos and wallabies

ii The Polyprotodonta comprises the American species and the Australasian carnivores which have a battery of pointed teeth in their rather elongated snouts.

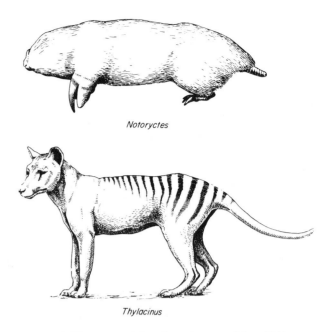

Notoryctes

Thylacinus

Fig. 10.3 The marsupial mole and marsupial wolf (drawn to different scales).

The Australasian herbivores have fewer premolars, the canine may be lacking, and, on each side, there are three incisors in the upper jaw and a large procumbent one in the lower jaw. In the Caenolestidae the first incisors in the lower jaw are large and procumbent but the more posterior incisors are retained—they are sometimes referred to in the literature by the unfortunate term paucituberculate. Foot structure also varies. The American genera and the Australasian dasyurids have five separate subequal digits on the hindfeet. In contrast, the Peramelidae together with all the Diprotodonta have the first digit on the hindfoot reduced to a nubbin, and the second and third digits partially fused to give a compound structure of about the same size as the fifth digit. The fourth digit can be larger, and is especially so in the saltatorial species.

iii The extensive adaptive radiation, which has produced many genera closely paralleling placental genera, is well known and summarized in Fig. 10.4. There are, however, no large, cursorial marsupial herbivores and the niches that are elsewhere occupied by the ungulates are occupied in Australasia by the large saltatorial kangaroos. Similarly, no easy placental parallels can be suggested for the so-called giant wombats—fossils such as *Diprotodon*. It is especially worth noting that during

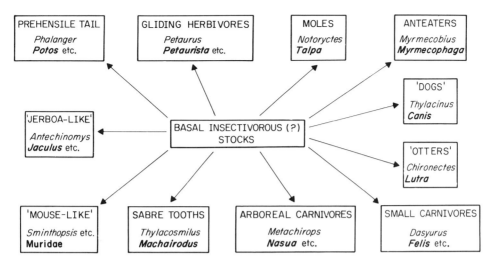

Fig. 10.4 A summary of the parallels between marsupials and placental genera.

much of the Tertiary, when the herbivores of South America were the indigenous Litopterna, Noto-ungulata, etc., the carnivores were marsupials—the Borhyaenidae.

Infraclass Eutheria—placentals

Various suggestions have been made about the inter-relationships of the orders of extinct and living placentals. The insectivores are undoubtedly close to the basal stock and the Primates retain many generalized characteristics. The Carnivora and the ungulate orders are reasonably associated within a cohort Ferungulata, but the affinities of various other orders are less clear. The rodents have been ascribed, alongside lagomorphs, to a separate cohort, the Glires, or considered to be a lineage derived from early Primates. In all orders there are numerous examples of parallel evolution in time, with different generic lineages filling the ecological niches currently occupied by living forms.

ORDER INSECTIVORA

The heterogeneity of extant insectivore genera is shown by Fig. 10.5. They are remnants of an ancient lineage and a fossil genus *Deltatheridium* occurs in the Upper Cretaceous of Mongolia. Many of their characteristics are primitive. The dental formula is $\frac{3}{3}:\frac{1}{1}:\frac{4}{4}:\frac{3-4}{3-4}$. The cranium has a generalized or primitive appearance; the

tympanic bone can be either free or included in an endotympanic bulla, and the testicles can be intra-abdominal, inguinal or within a prepenial scrotum recalling those of marsupials. Adaptations for a fossorial subterranean life occur independently in the families Chrysochloridae (golden moles) and Talpidae, and a hedgehog-like appearance occurs in both the Tenrecidae and Erinaceidae. Various genera such as *Potamogale* and *Galemys* are semi-aquatic. The close affinities between this order and the Primates are reflected in the varying taxonomic position to which the Tupaiidae, tree-shrews, are ascribed. Some authors place them in the Insectivora, others see them as Primates.

ORDER CHIROPTERA

The bats are the only mammals which fly. Various genera such as *Petaurus* amongst the Marsupialia, or *Glaucomys*, *Petaurista* and *Cynocephalus* amongst the Eutheria, glide. The bats have extensive alar membranes or patagia between the body, the limbs and their elongated digits. These membranes have three anatomical divisions. The propatagium is anterior to the forelimb and extends between the shoulders, the humerus, the radius and the base of the digits. The plagiopatagium is situated between the limbs and extends from the flanks to the ends of digits II to V. The uropatagium lies between the hindlimbs and the tail. The thumb is always free, as are the digits of the

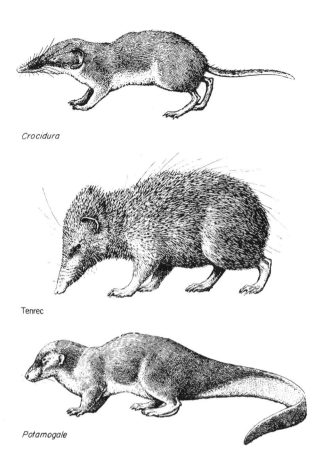

Crocidura

Tenrec

Potamogale

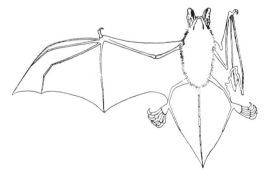

Fig. 10.5 Three insectivore types and an outline of a microchiropteran showing the components of the patagium.

hindlimb, and some bats have a sucker on their forelimbs. There are two principal sub-orders. The Megachiroptera includes forms like *Pteropus* the frugivorous flying-fox, and only one genus—*Rousettus*—uses echolocation. The Microchiroptera (Fig. 10.5) are

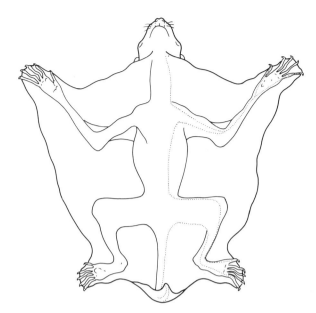

Fig. 10.6 Ventral view of *Cynocephalus* the 'flying colugo'.

predominantly insectivorous, echolocation is widely practised, and they are also sensitive to the Döppler shifts associated with the noise of flying insects. They are crepuscular or nocturnal and go into temporary 'hibernation' during the day.

ORDER DERMOPTERA

The genus *Cynocephalus* is of questionable affinities. It superficially resembles the 'Flying Squirrels' as it has a well-developed gliding membrane (Fig. 10.6). Its cheek teeth have affinities with those of some insectivores but the pectinate lower incisors are unmistakable. Colloquially referred to as the 'Flying Colugo' or 'Flying Lemur' it neither flies nor is it a lemur.

ORDER PRIMATES

i The characteristics of Primates are in large part adaptations to an arboreal life. The first digit is opposable, there is a voluminous cranium in which the fully ossified orbits look to the front, and the movements of the jaw are essentially vertical. Two major groupings are distinguished. The Strepsirhini, including the lemurs, lorises and galagos amongst modern forms, have a moist philtrum joining the moist rhinarium to the upper lip. This is lacking in the Haplorhini—tarsiers, monkeys, apes and men—the last three of which, at least, have menstrual cycles rather than oestrous cycles. An alternative taxonomic system involves grouping the tarsiers and lemuroid forms together in the Prosimii.

ii The tree-shrews are placed in the Strepsirhini by some authors and the other members of this grouping exhibit some considerable adaptive radiation (Fig. 10.7). Besides the well-known ring-tailed lemur there are ruffed lemurs, fruit-eating lemurs, and the rather different woolly lemurs of which the indri is the largest. All are restricted to Madagascar today but fossil families such as the Plesiadapidae and Adapidae were widespread prior to the Oligocene period. *Daubentonia*, the aye aye, also inhabits Madagascar, has two pairs of chisel-like gnawing teeth at the front and long, thin digits on its hand. Longest and thinnest of all is the middle finger which is used for combing its hair and reputedly for extricating insects from cracks, etc.

iii The lorises and galagos are more widespread. *Loris*

Daubentonia

Nycticebus

Lemur variegatus

Indri indri

Fig. 10.7 Some strepsirhine primates.

occurs in Ceylon and southern India; *Nycticebus*, the slow loris, in the Philippines and south Asia. Like *Perodicticus*, the potto, they are very slow-moving but the potto is unique in having bare vertebral spines, that it uses in defence, exposed outside its skin. In contrast the galagos of southern and equatorial Africa are nimble and their long, slender digits bear fleshy pads that facilitate rapid climbing. Like owls the galagos can rotate their heads to look backwards.

iv The Haplorhini includes four sub-divisions—Tarsioidea, Ceboidea, Cercopithecoidea and Hominoidea. The tarsiers are nocturnal and insectivorous. The pupils of their large, close-set eyes contract to a narrow slit in bright sunlight but during the night they expand to cover nearly the entire extent of the eye. The name Tarsier refers to the very long tarsus. Other noteworthy characters are the large ears, and the nails, rather than claws, on fingers and toes.

v The Ceboidea are the South American monkeys and near-monkeys such as the marmosets. Their alternative name—platyrrhines—refers to the large, oval nares which are directed laterally. The marmosets and tamarins possess hooked claws except in the case of the great toe which bears a true nail. Most of the monkeys possess prehensile tails. They include *Aotes*, the Douroucouli; *Cacajao*, the Uakari; *Pithecia*, the Saki; *Alouatta*, the Howler Monkey; *Cebus*, the Capuchin Monkey (Fig. 10.8); *Saimiri*, the Squirrel Monkey, and *Ateles*, the Spider Monkey. Their relationship to the Cercopithecoidea or Old World Monkeys remains unsolved.

vi The cercopithecoids include the macaques, guénons, langurs, mangabeys and baboons (Fig. 10.9 and 10.10). The macaques have cheek pouches in which they store food for short periods. There are callous patches on the buttocks which, as in baboons, for example, are often strikingly coloured. *Macaca mulatta*, the rhesus monkey, is perhaps the best known but others include *M. sylvanus*, the Barbary Ape; the pig-tailed macaque of India, Malaya and China; the lion-tailed macaque of Burma, etc., and the colourful Toque Macaque of Ceylon.

Members of the African genus *Cercopithecus* are known collectively as guénons—thought to mean 'maker of faces'. Some 80 species and races are recognized including the Talapoin, Green Monkey, Diana Monkey, Mona Monkey and Vervet. Also inhabiting Africa are the mangabeys—*Cercocebus*—which possess cheek pouches analogous to those of the oriental macaques. Amongst the better-known species are the Sooty, White-crowned and Crested Mangabeys. A

Fig. 10.9 A patas monkey, *Cercopithecus*.

Fig. 10.10 *Comopithecus hamadryas*, the hamadryas baboon.

Fig. 10.8 The capuchin monkey, a ceboid platyrrhine.

further African assemblage are the baboons. Their 'dog-like' skulls are unmistakable (Fig. 10.11) and in life they favour rocky country, eating fruits, roots, reptiles and insects. Much controversy has surrounded the taxonomic relationships of all Cercopithecidae and the animals currently ascribed to genera such as *Papio*, *Comopithecus* and *Mandrillus* have previously been associated with various other Old World forms in a variety of genera. The social life of all such monkeys is closely related to the female menstrual cycle. The dominant males tend to have access to the female closest to ovulation and less dominant ones to females further removed from ovulation. The dominant male and his consort maintain a social space around themselves.

The langurs or leaf monkeys, *Presbytis*, are the common monkeys of the orient. In broad terms they resemble the African guénons. Their long, straight tails facilitate the speedy movement through the trees and they can be distinguished from other genera by their shorter first digits and the absence of cheek pouches.

vii Finally there are the great apes and men. Our knowledge of the evolution of humans improves year by year. In broad terms the earliest forms of possible ape are *Parapithecus* and *Propliopithecus* from the Egyptian Oligocene deposits. The more recent Dryopithecinae,

that are known from widespread fossils of Miocene and Pliocene age, probably include the ancestors of all modern great apes and men. *Ramapithecus*, originally known from fragmentary material, is a near-hominid and the increasing number of hominid remains of pre-, or early, Pleistocene age demonstrate that human origins lie in the events of post-Miocene time. *Homo habilis* (Fig. 10.12) was clearly contemporary with *Australopithecus* and the Pleistocene saw a series of species of men—*H. habilis*, *H. erectus*, *H. sapiens*—that parallel the series of genera and species in other orders during the same period (see, for example, the proboscideans).

SUPER-ORDER CARNIVORA

i The carnivores of the early part of the Age of Mammals are now ascribed to a separate order—the Deltatheridia. The modern forms of the super-order Carnivora fall into two main groupings, the orders Fissipeda and Pinnepedia, comprising the terrestrial and aquatic carnivores respectively. The terrestrial

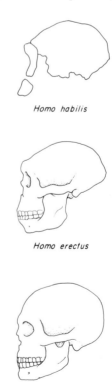

Homo habilis

Homo erectus

Homo sapiens

Fig. 10.12 The skulls of three species of *Homo*.

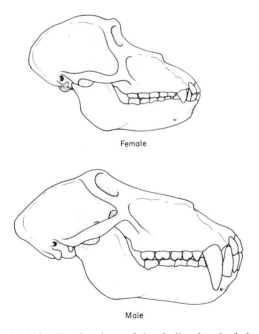

Female

Male

Fig. 10.11 Outline drawings of the skulls of male (below) and female baboons.

carnivores can be further associated into two lineages both of which evolved from small, long-bodied, plantigrade forms. The first, the Canoidea (Fig. 10.13), includes the Mustelidae, stoats and ermines; Procyonidae, coatis and kinkajoo; Canidae, wolves and dogs, and the Ursidae, bears. The bears are a post-Miocene lineage derived from the canid stock and during the Miocene a variety of 'dog-bears' and 'bear-dogs' existed. They are canids that are adapted to an omnivorous diet and can be identified by the numerous tubercles on their cheek teeth.

ii The Feloidea represent a somewhat separate lineage in which the Viverridae occupy the place of the Mustelidae in the Canoidea. Besides the civets, genets, and cats the association includes the hyaenas whose reduced dentition is paralleled by that of various fossil genera. In a similar manner sabre-toothed forms like *Smilodon* and *Machairodus* have evolved independently on a number of occasions (cf. the sabre-toothed marsupial *Thylacosmilus*).

iii The aquatic, and principally marine, Pinnipedia, comprises three families (Fig. 10.14). Both the Otariidae, sea lions, and the Odobaenidae, walruses, can rotate their hindlimbs forward for locomotion on land. They differ, however, in that the Otariidae have both external ears and a scrotum whilst these are absent from

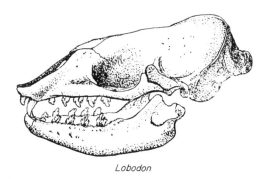
Lobodon

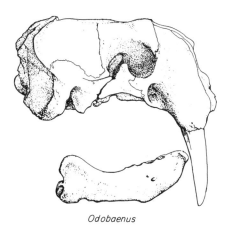
Odobaenus

Fig. 10.14 The skulls of *Lobodon* and the walrus *Odobaenus*.

walruses. The family Phocidae, seals, have their hindlimbs constantly directed backwards. A number of genera have a secondarily homodont or reduced dentition, but phocids such as *Lobodon* (Fig. 10.14) have a quite unmistakable dentition. The nasal cavity of *Cystophorus cristatus* is dilatable and *Mirounga leonina*, the elephant seal, can reach 6.5 metres in length. Pinniped origins may lie amongst the Miocene canoid stock.

ORDER CETACEA

i Primeval whales, Archeoceti, are quite widely represented as fossils but their precise origins are unknown although various authors have postulated a divergence from the Deltatheridia. They have very large brains with highly gyrencephalic hemispheres (Fig. 10.15). The Mysticeti appear to be anosmic. An outstanding feature of their circulatory system is the rete mirabile, or

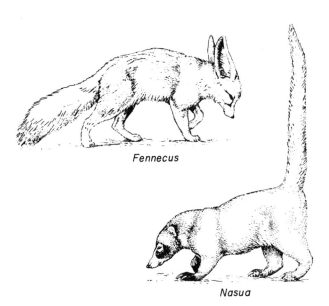
Fennecus

Nasua

Fig. 10.13 A fennec fox and a coati, two contrasting canoids.

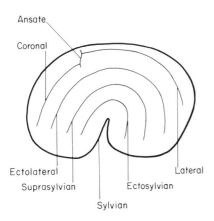

Fig. 10.15 Diagram of a developing whale brain showing the sulci characteristic of carnivores and ungulates. The situation subsequently becomes obscured. (After Hammelbo.)

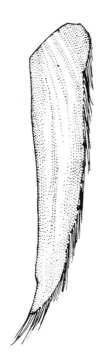

Fig. 10.16 A baleen plate of a whale.

capillary bed, within the arterial system. Such rete occur amongst the pinnipeds but are more extensive in whales. They seem to function as elastic reservoirs which accommodate and control the flow of arterial blood to the brain during diving.

ii Normally the sea constantly cools the cetacean, keeping its skin temperature close to that of the surrounding water. When it dives deep into the colder layers of the ocean its circulation is automatically reduced in proportion to depth. Blood flows more slowly through the sea-cooled peripheral vessels, thereby conserving heat. However, a flow of blood to the brain is essential but the method by which this is maintained is not fully understood. Although the brain and its blood-vessels and nerves are maintained at normal working temperature the heart beat drops during a long dive to about half the normal rate at the surface. In the killer whale this reduction is from 60 to 30 beats per minute but in Dall's porpoise and some other small dolphins the drop can be from 120 to 15 beats per minute. It is worth noting that similar falls to as low as 10% of the surface rate occur in beavers and hippopotamuses.

iii Two principal taxonomic groupings typify modern whales. These are the Odontoceti or toothed whales, and the Mysticeti or whale bone whales (Fig. 10.16). The first group includes the dolphins, porpoises, grampus and beluga, etc., the second, the Blue Whale, Humpback, etc. These have teeth during early development which are later lost and replaced by the elaborate baleen plates used for filtering krill such as *Euphausia*. The animals can weigh up to 100–130 tons.

SOUTH AMERICAN UNGULATE ORDERS

i The order Condylarthra is widely seen as including the basal stocks from which modern ungulate orders evolved. Genera which entered South America, prior to its isolation for much of the Tertiary, are also seen as the origin of the various indigenous fossil orders peculiar to that continent. These include some genera with unusual combinations of characters whose life forms are not easily imagined. Four orders are well represented by fossils and, of these, it is the Notoungulata that includes the greatest diversity. Historically, although Darwin brought home a specimen of *Toxodon*, and Lyddeker was well acquainted with the diversity involved, it was Ameghino who described the full multiplicity of form. Unfortunately he thought he was dealing with genera that were ancestral to all other orders and epitomized this in the generic names he created which provide a roll-call of nineteenth century zoologists, *Riccardolyddekeria, Ernestohaeckelia, Amilnedwardsia*, etc.

ii The order Litopterna comprises two families—the Proterotheriidae and Macraucheniidae. The genera of the former provide a striking parallel with the evolution of horses. *Diadiaphorus* is a rather tapiroid genus. The related *Proterotherium* had a longer facial region and longer limbs that suggest a more equid-like gait. This was carried even further in the contemporaneous *Thoatherium* which was totally monodactylous, with splint bones that were more reduced than those in horses. The Macraucheniidae is a distinct lineage and rather separate from the foregoing. The Oligocene *Theosodon* had proportions similar to those of llamas. There was a long neck and the nasal bones were short suggesting that it may have had a short proboscis. The nostrils were even further back in *Scalibrinitherium* and the condition reached its apogee in *Macrauchenia*. In this genus the long limbs, short neck and position of the nostrils high up on top of the cranium due to a foreshortening of the nasalia, give a very unusual appearance that has been interpreted as reflecting the presence of a proboscis or, alternatively, as reflecting a life underwater.

iii The Notoungulata includes a number of families. The Henricosborniidae, Arctostylopidae and Notostylopidae are somewhat primitive. The Typotheria and Hegetotheria (Fig. 10.17) are two distinct associations of small, possibly saltatorial, rodent-like forms distin-guished from each other by the differing form of their auditory region. The families Toxodontidae, Isotemnidae, Leontiniidae and Homalodotheriidae, usually united as the Toxodonta, include genera of larger body size. The Lower Eocene genus *Thomashuxleya* provides a possible ancestral form and the Isotemnidae and Homalodotheriidae are of particular interest as they had claws not hooves. In this they provide a striking parallel with the chalicotheres amongst fossil Perissodactyla (q.v.).

iv Other genera of large size are included in the order Astrapotheria. These were thought by Ameghino to be South American Amblypoda, early Tertiary gravigrade forms. However, the astrapotheres appear to be a totally indigenous evolutionary lineage of which the small-bodied *Trigonostylops* is perhaps an ancestral form.

v Finally, the Pyrotheria is one of the smallest but most distinctive and isolated of all eutherian orders. The type genus *Pyrotherium* from the Oligocene deposits of Patagonia is a very large animal, of gravigrade form, with very short nasal bones that again suggest the presence, in life, of a prehensile proboscis.

VARIOUS ARCHAIC ORDERS

During the Paleocene, Eocene and Oligocene a number of huge, herbivorous animals existed. These comprise at least three distinct orders. The Embrithopoda (10.18) had an awesome appearance and a uniform series of hypsodont teeth. The Pantodonta were a more heterogeneous assemblage, and the Dinocerata, again large, had paired bony protuberances on the skull. It was these last two orders that used to be united as the Amblypoda.

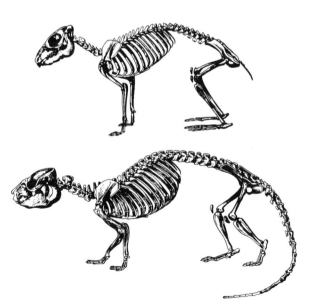

Fig. 10.17 The skeletons of a hegetothere (above) and a typothere (below). Small fossil notoungulates from South America. (After Scott.)

Fig. 10.18 A reconstruction of *Arsinotherium zitteli*, an Oligocene embrithopod from Egypt.

ORDER ARTIODACTYLA

i   The Artiodactyla are the even-toed ungulates, the axis of whose leg passes through digits III and IV. Modern forms fall fairly easily into three principal assemblages. The Suiformes includes pigs, peccaries and hippopotamuses; the Tylopoda (Fig. 10.19) constitutes a lineage that has been distinct since late Eocene times and is represented today by the camels, llamas, vicunas, etc., and the Ruminantia (Fig. 10.20) includes the deer

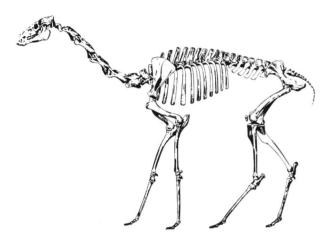

Fig. 10.19 The skeleton of *Oxydactylus*, a North American giraffe-necked camelid.

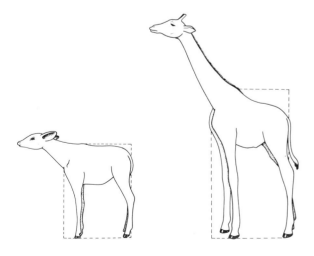

Fig. 10.20 Silhouettes of an okapi and a giraffe showing the differing proportions. (After Grassé.)

(Cervoidea), the giraffes and okapi (Giraffoidea) and the antelopes, goat-antelopes, cattle, sheep and goats (Bovoidea).

ii   This in no way gives any impression of the multiplicity of forms that have existed in the past. Large hippo-like forms occurred in the Entelodontinae, Anthraco-theriidae and Anoplotheriidae, which are usually placed in the Suiformes. Complementarily, the appearance of the suiform caenotheres paralleled modern rodents, or the typotheres and hegetotheres of South America. The Tylopoda included a number of giraffe-necked genera (e.g. *Alticamelus*) in North America during the Miocene and Pliocene. The Ruminantia, derived from small, antler-less, tusked Hypertragulinae of the Oligocene, includes many bizarre fossils. The Dromomerycinae, a sub-family of near-deer, included forms with three large horn cores. The giraffoid *Sivatherium*, together with related genera, was not only amongst the largest of known terrestrial animals but also had *two* pairs of horns, the anterior pair which were borne on the frontalia being conical, the more posterior pair borne on the parietalia being palmate. Many extinct bovoid species are also known from Pleistocene deposits and the precise interrelationships of living and fossil forms are subject to controversy.

iii   There is great variation in body size amongst the various genera of both deer and antelopes. The former vary in size from the tiny mouse deer to the extinct Irish Elk, and the latter from the tiny Imperial Antelope, comparable in size with large rodents, to the far larger Greater Kudu and Wildebeest. Many species have rather precise habitat preferences. Thus the Sittatunga, with its splayed digits, lives in or near water. The reed bucks and water bucks live in close proximity to water, as do various kobs and lechwes. Bush bucks are more closely associated with scrub, and the bongo exclusively with the dense forests of Zaïre. Species like the kudu inhabit hilly regions and various species like the Steinboks, oryxes and addaxes inhabit desert regions. Reference has already been made (page 134) to the various goat-antelopes that inhabit the himalayas, alps, Siberian salt-flats, etc.

ORDER PERISSODACTYLA

i   As with the Artiodactyla the modern world fauna gives a very scanty impression of the temporo-spatial diversity within the Perissodactyla. Modern genera are separated into the hippomorph and tapiromorph lineages on the basis of the W-shaped ectoloph on equid molars. The tapiromorphs include the tapirs themselves and the rhinoceroses. A plethora of fossil forms have been described.

ii Amongst the hippomorphs the genera ancestral to the horses are well known (Fig. 10.21). The bronto-

therioids are more bizarre forms (Fig. 10.22) which first drew attention to the great series of rich faunas in the White River deposits of America. However, the chalicotheres are the most unusual of all. Paralleled by the homalodotheres of South America, these were forms which combined bunodont, or bunolophodont, teeth with claws, which, in the genus *Moropus*, were retractile.

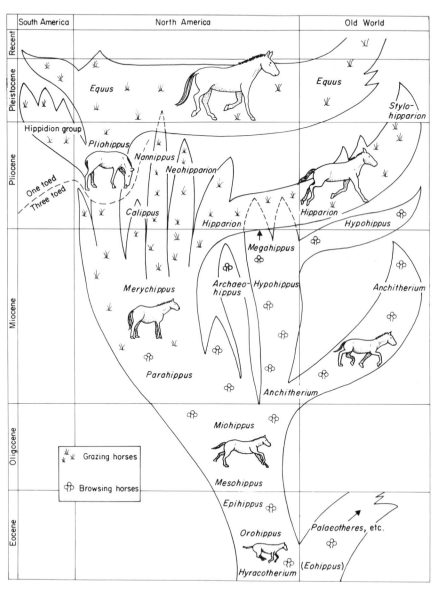

Fig. 10.21 The lineages of the horse family. (After Simpson.)

Fig. 10.22 *Brontotherium* from the Lower Oligocene of North America. (After Osborn.)

When first discovered in the early nineteenth century the claws and teeth were thought to belong to members of quite different orders.

iii The tapiromorph grouping also includes a number of unexpected fossil forms. The cursorial rhinoceroses included some genera comparable with the contemporaneous ancestral horses but others had long giraffe-like necks. In contrast, the amynodont rhinoceroses of the Upper Eocene and Oligocene were extremely heavily built animals, recalling hippopotamuses, and their short nasal bones elicit similar thoughts about the presence of a proboscis to those elicited by the macraucheniid litopterns.

ORDER PROBOSCIDEA

The unique and outstanding characters of the living elephants need no description. The large size of many fossil genera has also long drawn attention to them. They are generally supposed to originate in the moeritherioid genera which were rather 'hippopotamus-like' forms that were first described from the Oligocene Fayum beds of Egypt. The mastodon lineages of the later Tertiary are known in great detail. In broad terms they comprise a series of forms whose teeth show a gradual transition from the trituberculate appearance of the moeritherioids to trilophodont, tetralophodont and pentalophodont conditions (Fig. 10.23). In the genus *Stegolophodon*, of Upper Miocene to Quaternary age, the number of lamellae is even further increased and it is from some such forms that the elephants, with their many, high, transverse tooth ridges, are derived. The

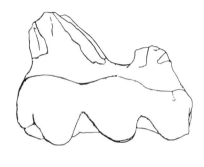

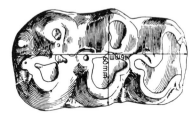

Fig. 10.23 Molar tooth of *Trilophodon*, a genus of mastodon. (After Osborn.)

mastodons survived until some 2000 years ago in America, and the elephant lineage involves a series of genera and species in the early, middle and late Quaternary interglacial deposits of the Old World.

ORDER TUBULIDENTATA

The aardvark (Fig. 10.24) is a digitigrade animal with secondarily homodont teeth, recalling those of

Fig. 10.24 The aardvark, *Orycteropus*, a tubulidentate.

Fig. 10.25 A hyrax. A member of yet another group that exhibits convergence with the rodents.

armadillos. Olfaction is a prominent sensory modality and the tongue is very sensitive. The distribution of *Orycteropus* coincides with that of African Isoptera. Its powers of digging to catch termites are considerable. The galleries which it digs can descend to a depth of three metres.

ORDER HYRACOIDEA

The hyrax (Fig. 10.25) is comparable in both appearance and habits to rodents and lagomorphs. There are prominent incisors, a diastema, and well-developed cheek teeth. As in the rodents and other herbivores the masseter muscle is also well developed.

ORDER SIRENIA

The sirenians, dugongs or manatees are aquatic mammals with a quite characteristic appearance. The anterior limbs form flippers and the head bears an enormous upper lip. They browse on aquatic vegetation and have been used in Guyana for many years to keep such vegetation from becoming too dense.

Fig. 10.26 The pangolin genus *Manis*. An Old World group that shows convergence with the South American armadillos.

ORDER PHOLIDOTA

The pangolins (Fig. 10.26) of the Old World tropics were formerly classed together with the Xenarthra as Edentata. It is now thought that their insectivorous habits and scaly appearance are the result of convergent evolution and that they have no close affinities with the South American forms. Teeth are totally absent and the tongue is very long and protrusible.

ORDER XENARTHRA

The anteaters, armadillos and sloths are all specialized South American mammals whose close affinities are questionable. They can nevertheless be conveniently placed together as Xenarthra and separated into three sub-orders.

The sub-order Tardigrada comprises the three genera of sloths, *Choloepus*, *Scaeopus* and *Bradypus*, whose unusual number of cervical vertebrae has already been remarked upon. The skulls are characteristic and with their secondarily homodont dentition, consisting of four or five peglike teeth, are quite unmistakable (Fig. 10.27). The Vermilingua, consisting of the three genera of anteaters, *Myrmecophaga*, *Tamandua* and *Cyclopes*, are also unmistakable (Fig. 10.28). The long, tubular form of the skull associated with insect-eating habits parallels

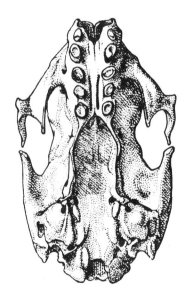

Fig. 10.27 A palatine view of a sloth skull. Note the secondarily homodont peglike teeth.

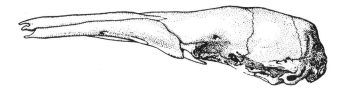

Fig. 10.28 The skull, less lower jaw, of an anteater.

that of the echidna and marsupial anteaters. The Loricata or armadillos retain a more usual skull form but here again the dentition is reduced to some seven peglike teeth in each half jaw. Their embryology is unique in that the egg, following a normal early development, subsequently gives rise to eight or twelve embryos—a condition known as poly-embryony.

ORDER LAGOMORPHA

The pikas, rabbits, hares, cottontails and their allies have two incisors on each side of their upper jaw—a condition known as duplicidentate—whereas the rodents have only one such incisor, are simplicidentate, but the two orders are frequently united in a single cohort— the Glires. The pikas resemble marmots in their overall appearance, are widespread, and can be found from Tibet to America. The Rocky Mountain species *Ochotona princeps* is well known for its habit of collecting and laying out to dry large quantities of vegetable material prior to hibernation. The remaining genera vary considerably in size. Various arctic hares, such as *Lepus arcticus*, are amongst the largest, whilst the pygmy species *Romerolagus nelsoni* from Mount Popocatapetl, and the Idaho pygmy hare, *Brachylagus*, are the smallest. Strictly speaking, the term rabbit should be restricted to *Oryctolagus cuniculus*, as the various cottontails and swamp rabbits of America are *Sylvilagus*.

ORDER RODENTIA

i As mentioned above, the principal formal distinction between rodents and lagomorphs is the presence of one and two pairs of incisors, respectively, in the upper jaw. Like the bovoid artiodactyls the rodents have undergone extensive adaptive radiations in the more recent phases of the Cenozoic Era and the precise relationships of many families are, therefore, not clear. However, three major associations can be distinguished and comprise the squirrel-like, the mouse-like, and the porcupine-like genera—Sciuromorpha, Myomorpha and Hystricomorpha, respectively. These have differing

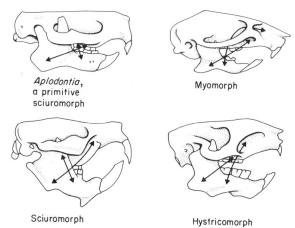

Fig. 10.29 The differing disposition of the masseter muscle in rodents. In *Aplodontia* it originates from the zygomatic arch. In myomorphs the deep component traverses the orbit to insert on the facial region. In so-called advanced sciuromorphs the superficial component is inserted on the face. In hystricomorphs there is a foramen for the deep component. (After Romer.)

degrees of complexity of their cheek teeth and masseter muscles (Fig. 10.29).

ii The most ancient fossil genera had simple molars and there has subsequently been a gradual increase in complexity leading to the pentalophodont condition of Hystricomorpha in which there are five prominent surface ridges. The fossil family Ischyromyidae provides a link between the most archaic and more recent forms but amongst living species it is the sewellel which retains the most archaic anatomy, and its family, the Aplodontidae, is sometimes separated from all the rest.

iii The Sciuromorpha includes, besides the squirrels, the marmots, flying squirrels and scaly-tailed squirrels (Anomalurinae). The squirrels themselves vary in size considerably, from the giant Asiatic *Ratufa*, to the tiny flame squirrels of South America, or *Nannosciurus*, the naked squirrel of Malaysia.

iv The forms aggregated within the Myomorpha include all the mouse- or jerboa-like families together with rather aberrant aquatic genera from Australasia such as *Hydromys*.

v The Hystricomorpha includes both the Old World and New World porcupines, *Hystrix* and *Erethrizon*, which are only rather distantly related; the beavers, and a widely divergent series of South American forms. Some Hystricomorpha entered that region in the Miocene or Pliocene during the early phases of the establishment of the isthmus of Panama. They then underwent an extensive adaptive radiation giving rise to the agoutis, coypus, pacas and capybaras whose way of life compares in many respects with those of various artiodactyls.

FURTHER READING

ANDERSEN H.T. (ed.) (1969) *The biology of marine mammals.* Academic Press, London.

GUNDERSON H.L. (1976) *Mammalogy.* McGraw Hill, New York.

MORRIS D. (1965) *The mammals.* Hodder and Stoughton, London.

TROUGHTON E. (1965) *Furred animals of Australia.* Angus and Robertson, London.

TYNDALE-BISCOE H. (1973) *Life of marsupials.* Edward Arnold, London.

VAUGHAN T.A. (1972) *Mammalogy.* W.B. Saunders and Co., Philadelphia.

WALKER E.P. (ed.) (1964) *Mammals of the World*, 3 vols. Johns Hopkins University Press, London.

YOUNG J.Z. (1975) *The life of mammals*, (2nd ed.). Oxford University Press, Oxford.

Taxonomic Index

Page numbers in italic indicate pages which include references to text and illustrations or solely to illustrations.

Subject Index

Page numbers in italic indicate pages which include references to text and illustrations or solely to illustrations.